T0211909

Lifelong Machine Learning

Second Edition

Synthesis Lectures on Artificial Intelligence and Machine Learning

Editors
Ronald J. Brachman, *Jacobs Technion-Cornell Institute at Cornell Tech*
Peter Stone, *University of Texas at Austin*

Markov Logic: An Interface Layer for Artificial Intelligence
Pedro Domingos and Daniel Lowd
2009

Introduction to Semi-Supervised Learning
XiaojinZhu and Andrew B.Goldberg
2009

Action Programming Languages
Michael Thielscher
2008

Representation Discovery using Harmonic Analysis
Sridhar Mahadevan
2008

Essentials of Game Theory: A Concise Multidisciplinary Introduction
Kevin Leyton-Brown and Yoav Shoham
2008

A Concise Introduction to Multiagent Systems and Distributed Artificial Intelligence
Nikos Vlassis
2007

Intelligent Autonomous Robotics: A Robot Soccer Case Study
Peter Stone
2007

Lifelong Machine Learning, Second Edition
Zhiyuan Chen and Bing Liu

ISBN: 978-3-031-00453-7   paperback
ISBN: 978-3-031-01581-6   ebook
ISBN: 978-3-031-02709-3   epub
ISBN: 978-3-031-00026-3   hardcover

DOI 10.1007/978-3-031-01581-6

A Publication in the Springer series
SYNTHESIS LECTURES ON ARTIFICIAL INTELLIGENCE AND MACHINE LEARNING

Lecture #38
Series Editors: Ronald J. Brachman, *Jacobs Technion-Cornell Institute at Cornell Tech*
                Peter Stone, *University of Texas at Austin*
Series ISSN
Print 1939-4608   Electronic 1939-4616

Lifelong Machine Learning

Second Edition

Zhiyuan Chen
Google, Inc.

Bing Liu
University of Illinois at Chicago

SYNTHESIS LECTURES ON ARTIFICIAL INTELLIGENCE AND MACHINE LEARNING #38

ABSTRACT

Lifelong Machine Learning, Second Edition is an introduction to an advanced machine learning paradigm that continuously learns by accumulating past knowledge that it then uses in future learning and problem solving. In contrast, the current dominant machine learning paradigm learns in isolation: given a training dataset, it runs a machine learning algorithm on the dataset to produce a model that is then used in its intended application. It makes no attempt to retain the learned knowledge and use it in subsequent learning. Unlike this isolated system, humans learn effectively with only a few examples precisely because our learning is very knowledge-driven: the knowledge learned in the past helps us learn new things with little data or effort. Lifelong learning aims to emulate this capability, because without it, an AI system cannot be considered truly intelligent.

Research in lifelong learning has developed significantly in the relatively short time since the first edition of this book was published. The purpose of this second edition is to expand the definition of lifelong learning, update the content of several chapters, and add a new chapter about continual learning in deep neural networks—which has been actively researched over the past two or three years. A few chapters have also been reorganized to make each of them more coherent for the reader. Moreover, the authors want to propose a unified framework for the research area. Currently, there are several research topics in machine learning that are closely related to lifelong learning—most notably, multi-task learning, transfer learning, and meta-learning—because they also employ the idea of knowledge sharing and transfer. This book brings all these topics under one roof and discusses their similarities and differences. Its goal is to introduce this emerging machine learning paradigm and present a comprehensive survey and review of the important research results and latest ideas in the area. This book is thus suitable for students, researchers, and practitioners who are interested in machine learning, data mining, natural language processing, or pattern recognition. Lecturers can readily use the book for courses in any of these related fields.

KEYWORDS

lifelong machine learning; lifelong learning; continuous learning; continual learning; meta-learning, never-ending learning; multi-task learning; transfer learning

Zhiyuan dedicates this book to his wife, Vena Li, and his parents.

Bing dedicates this book to his wife, Yue He; his children,
Shelley and Kate; and his parents.

Contents

Preface

The purpose of writing this second edition is to extend the definition of lifelong learning, to update the content of several chapters, and to add a new chapter about *continual learning in deep neural networks*, which has been actively researched for the past two to three years. A few chapters are also reorganized to make each of them more coherent.

The project of writing this book started with a tutorial on *lifelong machine learning* that we gave at the 24th International Joint Conference on Artificial Intelligence (IJCAI) in 2015. At that time, we had worked on the topic for a while and published several papers in ICML, KDD, and ACL. When Morgan & Claypool Publishers contacted us about the possibility of developing a book on the topic, we were excited. We strongly believe that lifelong machine learning (or simply lifelong learning) is very important for the future of machine learning and artificial intelligence (AI). Note that lifelong learning is sometimes also called *continual learning* or *continuous learning* in the literature. Our original research interest in the topic stemmed from extensive application experiences in sentiment analysis (SA) in a start-up company several years ago. A typical SA project starts with a client who is interested in consumer opinions expressed in social media about their products or services and those of their competitors. There are two main analysis tasks that an SA system needs to do: (1) discover the entities (e.g., *iPhone*) and entity attributes/features (e.g., *battery life*) that people talked about in opinion documents such as online reviews and (2) determine whether the opinion about each entity or entity attribute is positive, negative, or neutral [Liu, 2012, 2015]. For example, from the sentence "*iPhone is really cool, but its battery life sucks,*" an SA system should discover that the author is (1) positive about *iPhone* and (2) negative about iPhone's *battery life*.

After working on many projects in many domains (which are types of products or services) for clients, we realized that there is a great deal of sharing of information across domains and projects. As we see more and more, new things get fewer and fewer. It is easy to see that sentiment words and expressions (such as *good, bad, poor, terrible,* and *cost an arm and a leg*) are shared across domains. There is also a great deal of sharing of entities and attributes. For example, every product has the attribute of *price*, most electronic products have *battery*, and many of them also have *screen*. It is silly not to exploit such sharing to significantly improve SA to make it much more accurate than without using such sharing but only working on each project and its data in isolation. The classic machine learning paradigm learns exactly in isolation. Given a dataset, a learning algorithm runs on the data to produce a model. The algorithm has no memory and thus is unable to use the previously learned knowledge. In order to exploit knowledge sharing, an SA system has to retain and accumulate the knowledge learned in the past and use it to help future learning and problem solving, which is exactly what *lifelong learning* aims to do.

It is not hard to imagine that this sharing of information or knowledge across domains and tasks is generally true in every field. It is particularly obvious in natural language processing because the meanings of words and phrases are basically the same across domains and tasks and so is the sentence syntax. No matter what subject matter we talk about, we use the same language, although each subject may use only a small subset of the words and phrases in a language. If that is not the case, it is doubtful that a natural language would have ever been developed by humans. Thus, lifelong learning is generally applicable, not just restricted to sentiment analysis.

The goal of this book is to introduce this emerging machine learning paradigm and to present a comprehensive survey and review of the important research results and latest ideas in the area. We also want to propose a unified framework for the research area. Currently, there are several research topics in machine learning that are closely related to lifelong learning, most notably, multi-task learning and transfer learning, because they also employ the idea of knowledge sharing and transfer. This book brings all these topics under one roof and discusses their similarities and differences. We see lifelong learning as an extension to these related paradigms. Through this book, we would also like to motivate and encourage researchers to work on lifelong learning. We believe it represents a major research direction for both machine learning and artificial intelligence for years to come. Without the capability of retaining and accumulating knowledge learned in the past, making inferences about it, and using the knowledge to help future learning and problem solving, achieving artificial general intelligence (AGI) is unlikely.

Two main principles have guided the writing of this book. First, it should contain strong motivations for conducting research in lifelong learning in order to encourage graduate students and researchers to work on lifelong learning problems. Second, the writing should be accessible to practitioners and upper-level undergraduate students who have basic knowledge of machine learning and data mining. Yet there should be sufficient in-depth materials for graduate students who plan to pursue Ph.D. degrees in the machine learning and/or data mining fields.

This book is thus suitable for students, researchers, and practitioners who are interested in machine learning, data mining, natural language processing, or pattern recognition. Lecturers can readily use the book in class for courses in any of these related fields.

Zhiyuan Chen and Bing Liu
August 2018

Acknowledgments

We would like to thank the current and former graduate students in our group and our collaborators: Geli Fei, Zhiqiang Gao, Estevam R. Hruschka Jr., Wenpeng Hu, Minlie Huang, Yongbing Huang, Doo Soon Kim, Huayi Li, Jian Li, Lifeng Liu, Qian Liu, Guangyi Lv, Sahisnu Mazumder, Arjun Mukherjee, Nianzu Ma, Lei Shu, Tao Huang, William Underwood, Hao Wang, Shuai Wang, Hu Xu, Yueshen Xu, Tim Yin, Tim Yuan, and Yuanlin Zhang, for their contributions of numerous research ideas and helpful discussions over the years. We are especially grateful to the two expert reviewers of the first edition, Eric Eaton and Matthew E. Taylor. Despite their busy schedules, they read the first draft of the book very carefully and gave us so many excellent comments and suggestions, which were not only insightful and comprehensive, but also detailed and very constructive. German I. Parisi reviewed Chapter 4 of this second edition and gave us many valuable comments. Their suggestions have helped us improve the book tremendously.

On the publication side, we thank the editors of Synthesis Lectures on Artificial Intelligence and Machine Learning, Ronald Brachman, William W. Cohen, and Peter Stone, for initiating this project. The President and CEO of Morgan & Claypool Publishers, Michael Morgan, and his staff, Christine Kiilerich, and C.L. Tondo have given us all kinds of help promptly whenever requested, for which we are very grateful.

Our greatest gratitude go to our own families. Zhiyuan Chen would like to thank his wife Vena Li and his parents. Bing Liu would like to thank his wife Yue, his children Shelley and Kate, and his parents. They have helped in so many ways.

The writing of this book was partially supported by two National Science Foundation (NSF) grants IIS-1407927 and IIS-1650900, an NCI grant R01CA192240, a research gift from Huawei Technologies, and a research gift from Robert Bosch GmbH. The content of the book is solely the responsibility of the authors and does not necessarily represent the official views of the NSF, NCI, Huawei, or Bosch. The Department of Computer Science at the University of Illinois at Chicago provided computing resources and a very supportive environment for this project. Working at Google has also given Zhiyuan Chen a broader perspective on machine learning.

Zhiyuan Chen and Bing Liu
August 2018

CHAPTER 1

Introduction

Machine learning (ML) has been instrumental for the advance of both data analysis and artificial intelligence (AI). The recent success of deep learning brought ML to a new height. ML algorithms have been applied in almost all areas of computer science, natural science, engineering, social sciences, and beyond. Practical applications are even more widespread. Without effective ML algorithms, many industries would not have existed or flourished, e.g., Internet commerce and Web search. However, the current ML paradigm is not without its weaknesses. In this chapter, we first discuss the classic ML paradigm and its shortcomings, and then introduce *Lifelong ML* (or simply *Lifelong Learning* (LL)) as an emerging and promising direction to overcome those shortcomings with the ultimate goal of building machines that learn like humans.

1.1 CLASSIC MACHINE LEARNING PARADIGM

The current dominant paradigm for ML is to run an ML algorithm on a given dataset to generate a model. The model is then applied in real-life performance tasks. This is true for both supervised learning and unsupervised learning. We call this paradigm *isolated learning* because it does not consider any other related information or the previously learned knowledge. The fundamental problem with this isolated learning paradigm is that it does not retain and accumulate knowledge learned in the past and use it in future learning. This is in sharp contrast to our human learning. We humans never learn in isolation or from scratch. We always retain the knowledge learned in the past and use it to help future learning and problem solving. Without the ability to accumulate and use the past knowledge, an ML algorithm typically needs a large number of training examples in order to learn effectively. The learning environments are typically static and closed. For supervised learning, labeling of training data is often done manually, which is very labor-intensive and time-consuming. Since the world is too complex with too many possible tasks, it is almost impossible to label a large number of examples for every possible task or application for an ML algorithm to learn. To make matters worse, everything around us also changes constantly, and the labeling thus needs to be done continually, which is a daunting task for humans. Even for unsupervised learning, collecting a large volume of data may not be possible in many cases.

In contrast, we humans learn quite differently. We accumulate and maintain the knowledge learned from previous tasks and use it seamlessly in learning new tasks and solving new problems. That is why whenever we encounter a new situation or problem, we may notice that many aspects of it are not really new because we have seen them in the past in some other contexts. When faced with a new problem or a new environment, we can adapt our past knowledge to deal with the new situation and also learn from it. Over time we learn more and more, and become more and more knowledgeable and more and more effective at learning. *Lifelong machine learning* or simply *lifelong learning* (LL) aims to imitate this human learning process and capability. This type of learning is quite natural because things around us are closely related and interconnected. Knowledge learned about some subjects can help us understand and learn some other subjects. For example, we humans do not need 1,000 positive online reviews and 1,000 negative online reviews of movies as an ML algorithm needs in order to build an accurate classifier to classify positive and negative reviews about a movie. In fact, for this task, without a single training example, we can already perform the classification task. How can that be? The reason is simple. It is because we have accumulated so much knowledge in the past about the language expressions that people use to praise or to criticize things, although none of those praises or criticisms may be in the form of online reviews. Interestingly, if we do not have such past knowledge, we humans are probably unable to manually build a good classifier even with 1,000 training positive reviews and 1,000 training negative reviews without spending an enormous amount of time. For example, if you have no knowledge of Arabic and someone gives you 2,000 labeled training reviews in Arabic and asks you to build a classifier manually, most probably you will not be able to do it without using a translator.

To make the case more general, we use natural language processing (NLP) as an example. It is easy to see the importance of LL to NLP for several reasons. First, words and phrases have almost the same meaning in all domains and all tasks. Second, sentences in every domain follow the same syntax or grammar. Third, almost all natural language processing problems are closely related to each other, which means that they are inter-connected and affect each other in some ways. The first two reasons ensure that the knowledge learned can be used across domains and tasks due to the sharing of the same expressions and meanings and the same syntax. That is why we humans do not need to re-learn the language (or to learn a new language) whenever we encounter a new application domain. For example, assume we have never studied psychology, and we want to study it now. We do not need to learn the language used in the psychology text except some new concepts in the psychology domain because everything about the language itself is the same as in any other domain or area. The third reason ensures that LL can be used across different types of tasks. For example, a named entity recognition (NER) system has learned that iPhone is a product or entity, and a data mining system has discovered that every product has a price and the adjective "expensive" describes the price attribute of an entity. Then, from the sentence *"The picture quality of iPhone is great, but it is quite expensive,"* we can safely extract "picture quality" as a feature or attribute of iPhone, and detect that "it" refers to iPhone not the picture quality with the help of those pieces of prior knowledge. Traditionally, these problems are solved separately in isolation, but they are all related and can help each other because the results from one problem can be useful to others. This situation is common for all NLP tasks. Note that we regard anything from unknown to known as a piece of knowledge. Thus, a learned model is a piece of knowledge and the results gained from applying the model are also knowledge, although they are different kinds of knowledge. For example, iPhone being an entity and picture quality being an attribute of iPhone are two pieces of knowledge.

Realizing and being able to exploit the sharing of words and expressions across domains and inter-connectedness of tasks are still insufficient. A large quantity of knowledge is often needed in order to help the new task learning effectively because the knowledge gained from one previous task may contain only a tiny bit or even no knowledge that is applicable to the new task (unless the two tasks are extremely similar). Thus, it is important to learn from a large number of diverse domains to accumulate a large amount of diverse knowledge. A future task can pick and choose the appropriate past knowledge to use to help its learning. As the world also changes constantly, the learning should thus be continuous or lifelong, which is what we humans do.

Although we used NLP as an example, the general idea is true for any other area because again things in the world are related and inter-connected. There is probably nothing that is completely unrelated to anything else. Thus, knowledge learned in the past in some domains can be applied in some other domains with similar contexts. The classic isolated learning paradigm is unable to perform such LL. As mentioned earlier, it is only suitable for narrow and restricted tasks in closed environments. It is also probably not sufficient for building an intelligent system that can learn continually to achieve close to the human level of intelligence. LL aims to make progress in this very direction. With the popularity of robots, intelligent personal assistants, and chatbots, LL is becoming increasingly important because these systems have to interact with humans and/or other systems, learn constantly in the process, and retain the knowledge learned in their interactions in the ever-changing environments to enable them to learn more and to function better over time.

1.2   MOTIVATING EXAMPLES

In the above, we motivated LL from the perspective of human learning and NLP. In this section, we use some concrete examples, i.e., sentiment analysis, self-driving cars, and chatbots, to further motivate LL. Our original motivation for studying LL actually stemmed from extensive application experiences in sentiment analysis (SA) in a start-up company several years ago. There are two main tasks that an SA system needs to perform: The first task is usually called aspect extraction, which discovers the entities (e.g., iPhone) and entity attributes/features (e.g., battery life) that people talked about in an opinion document such as an online review. These entities and entity attributes are commonly called *aspects* in SA. The second task is to determine whether an opinion about an aspect (entity or entity attribute) is positive, negative, or neutral [Liu, 2012, 2015]. For example, from the sentence "iPhone is really cool, but its battery life sucks," an SA system should discover that the author is positive about iPhone but negative about iPhone's battery life.

There are two main types of application scenarios. The first type is to analyze consumer opinions about one particular product or service (or a small number of products or services), e.g., iPhone or a particular hotel. This kind of application is highly focused and usually not very difficult. The second type is to analyze consumer opinions about a large number of products or services, e.g., opinions about all products sold on Amazon's or Best Buy's websites. Although compared to the first type, the second type is just a quantity change, in fact, it leads to a sea quality change because the techniques used for the first type may no longer be practical for the second type. Let us look at both the supervised and the unsupervised approaches to performing these tasks.

We first analyze the supervised approach. For the first type of application, it is reasonable to spend some time and effort to manually label a large amount of data for aspect extraction and sentiment classification. Note that these are very different tasks and require different kinds of labeling or annotations. With the labeled training data, we can experiment with different ML models, tune their parameters, and design different features in order to build good models for extraction and for classification. This approach is reasonable because we only need to work on opinions about one product or service. In the unsupervised approach, the common method is to use syntactic rules compiled by humans for aspect extraction. For sentiment classification, the common approach is to use a list sentiment words and phrases (e.g., good, bad, beautiful, bad, horrible, and awful) and syntactic analysis to decide sentiments. Although these methods are called unsupervised, they are not completely domain independent. In different domains, extraction rules could be different because people may have somewhat different ways to express opinions. For sentiment classification, a word may be positive in one domain or even in one particular context but negative in another domain. For example, for the word "quiet," the sentence "this car is very quiet" in the car domain is positive, but the sentence "this earphone is very quiet" is negative in the earphone domain. There are also other difficult issues [Liu et al., 2015b]. If we only need to deal with one or two domains (products or services), we can spend time to handcraft rules and identify those domain specific sentiments in order to produce accurate extraction and classification systems.

However, for the second type of application, these two approaches become problematic because they cannot scale up. Amazon.com probably sells hundreds of thousands, if not more, of different products. To label a large amount of data for each kind of product is a daunting task, not to mention new products are launched all the time. It is well known that labeled training data in one domain does not work well for another domain. Although transfer learning can help, it is inaccurate. Worse still, transfer learning usually requires the human user to provide similar source and target domains; otherwise, it can result in negative transfer, and generate poorer results. Although crowdsourcing may be used for labeling, the quality of the labeled data is an issue. More importantly, most products sold on the Web do not have a lot of reviews, which is insufficient for building accurate classifiers or extractors. For the unsupervised approach, the problem is the same. Every type of product is different. For a human to handcraft extraction rules and identify sentiment words with domain-specific sentiment polarities is also an almost impossible task.

Although the traditional approach is very difficult for the second type of application, it does not mean there is no possible solution. After working on many projects for clients in a start-up company, we realized that there are a significant amount of sharing of knowledge for both aspect extraction and sentiment classification across domains (or different types of products). As we see reviews of more and more products, new things get fewer and fewer. It is easy to notice that sentiment words and expressions (such as good, bad, poor, terrible, and cost an arm and a leg) are shared across domains. There is also a great deal of sharing of aspects (entities and attributes). For example, every product has the attribute of price, most electronic products have batteries, and many of them also have a screen. It is silly not to exploit such sharing to significantly improve SA to make it much more accurate than without using such sharing but only working on the reviews of each product in isolation.

This experience and intuition led us to try to find a systematic way to exploit the knowledge learned in the past. LL is the natural choice as it is a paradigm that learns continually, retains the knowledge learned in the past, and uses the accumulated knowledge to help future learning and problem solving. LL can be applied to both supervised and unsupervised learning approaches to SA. It can enable sentiment analysis to scale up to a very large number of domains. In the supervised approach, we no longer need a large number of labeled training examples. In many domains, no training data is needed at all because they may already be covered by some other/past domains and such similar past domains can be automatically discovered. In the unsupervised approach, it also enables the system to perform more accurate extraction and sentiment classification because of the shared knowledge. It is also possible to automatically discover those domain-specific sentiment polarities of words in a particular domain. We will see some of the techniques in this book.

Interestingly, this application of LL led to two critical problems, i.e., the correctness of knowledge and the applicability of knowledge. Before using a piece of past knowledge for a particular domain, we need to make sure that the piece of past knowledge is actually correct. If it is correct, we must also make sure that it is applicable to the current domain. Without dealing with these two problems, the results in the new domain can get worse. In the later part of the book, we will discuss some methods for solving these problems in both the supervised and unsupervised settings.

For self-driving cars, the situation is similar. There are again two basic approaches to learning to drive: rule-based approach and learning-based approach. In the rule-based approach, it is very hard to write rules to cover all possible scenarios on the road. The learning-based approach has a similar issue because the road environment is highly dynamic and complex. We use the perception system as an example. For the perception system to detect and recognize all kinds of objects on the road in order to predict potential hazards and dangerous situations, it is extremely hard to train a system based on labeled training data. It is highly desirable that the system can perform continuous learning during driving and in the process identify unseen objects and learn to recognize them, and also learn their behaviors and danger levels to the vehicle by making use of the past knowledge and the feedback from the surround environment. For example, when the car sees a black patch on the road that it has never seen before, it must first recognize that this is an unseen object and then incrementally learn to recognize it in the future, and to assess its danger level to the car. If the other cars have driven over it (environmental feedback), it means that the patch is not dangerous. In fact, on the road, the car can learn a great deal of knowledge from the cars before and after it. This learning process is thus self-supervised (with no external manual labeling of the data) and never ends. As time goes by, the car becomes more and more knowledgeable and smarter and smarter.

Finally, we use the development of chatbots to further motivate LL. In recent years, chatbots have become very popular due to their widespread applications in performing goal-oriented tasks (like assisting customers in buying products, booking flight tickets, etc.) and accompanying humans to get rid of stress via open-ended conversations. Numerous chatbots have been developed or are under development, and many researchers are also actively working on techniques for chatbots. However, there are still some major weaknesses with the current chatbots that limit the scope of their applications. One serious weakness of the current chatbots is that they cannot learn new knowledge during conversations, i.e., their knowledge is fixed beforehand and cannot be expanded or updated during the conversation process. This is very different from our human conversations. We human beings learn a great deal of knowledge in our conversations. We either learn from the utterances of others, or by asking others if we do not understand something. For example, whenever we encounter an unknown concept in a user question or utterance, we try to gather information about it and reason in our brain by accessing the knowledge in our long-term memory before answering the question or responding to the utterance. To gather information, we typically ask questions to the persons whom we are conversing with because acquiring new knowledge through interaction with others is a natural tendency of human beings. The newly acquired information or knowledge not only assists the current reasoning task, but also helps future reasoning. Thus, our knowledge grows over time. As time goes by, we become more and more knowledgeable and better and better at learning and conversing. Naturally, chatbots should have this LL or continuous learning capability. In Chapter 8, we will see an initial attempt to make chatbots learn during conversations.

1.3   A BRIEF HISTORY OF LIFELONG LEARNING

The concept of *lifelong learning* (LL) was proposed around 1995 in Thrun and Mitchell [1995]. Since then it has been pursued in several directions. We give a brief history of the LL research in each of the directions below.

1. *Lifelong Supervised Learning*. Thrun [1996b] first studied lifelong concept learning, where each previous or new task aims to recognize a particular concept or class using binary classification. Several LL techniques were proposed in the contexts of memory-based learning

and neutral networks. The neural network approach was improved in Silver and Mercer [1996, 2002] and Silver et al. [2015]. Ruvolo and Eaton [2013b] proposed an efficient lifelong learning algorithm (ELLA) to improve the multi-task learning (MTL) method in Kumar et al. [2012]. Here the learning tasks are independent of each other. Ruvolo and Eaton [2013a] also considered LL in an active task selection setting. Chen et al. [2015] proposed an LL technique in the context of Naïve Bayesian (NB) classification. A theoretical study of LL was done by Pentina and Lampert [2014] in the PAC-learning framework. Shu et al. [2017b] proposed a method to improve a conditional random fields (CRF) model during model application or testing. It is like *learning on the job*, which other existing models cannot do. Mazumder et al. [2018] worked along a similar line in the context of human-machine conversation to enable chatbots to continually learn new knowledge in the conversation process.

2. *Continual Learning in Deep Neural Networks.* In the past few years, due to the popularity of deep learning, many researchers studied the problem of continually learning a sequence of tasks in the deep learning context [Parisi et al., 2018a]. Note that LL is also called *continual learning* in the deep learning community. The main motivation of continual learning in deep learning is to deal with the problem of *catastrophic forgetting* when learning a series of tasks [McCloskey and Cohen, 1989]. The focus has been on incrementally learning each new task in the same neural network without causing the neural network to forget the models learned for the past tasks. Limited work has been done on how to leverage the previously learned knowledge to help learn the new task better. This is in contrast to the other LL methods, which emphasize leveraging the past knowledge to help new learning.

3. *Open-world Learning.* Traditional supervised learning makes the *closed-world assumption* that the classes of the test instances must have been seen in training [Bendale and Boult, 2015, Fei and Liu, 2016]. This is not suitable for learning in open and dynamic environments because in such an environment, there are always new things showing up. That is, in the model testing or application, some instances from unseen classes may appear. Open-world learning deals with this situation [Bendale and Boult, 2015, Fei et al., 2016, Shu et al., 2017a]. That is, an open-world learner must be able to build models that can detect unseen classes during testing or the model application process, and also learn the new classes incrementally based on the new classes and the old model.

4. *Lifelong Unsupervised Learning.* Papers in this area are mainly about lifelong topic modeling and lifelong information extraction. Chen and Liu [2014a,b] and Wang et al. [2016] proposed several lifelong topic modeling techniques that mine knowledge from topics produced from many previous tasks and use it to help generate better topics in the new task. Liu et al. [2016] also proposed an LL approach based on recommendation for information extraction in the context of opinion mining. Shu et al. [2016] proposed a lifelong relax-

ation labeling method to solve a unsupervised classification problem. These techniques are all based on meta-level mining, i.e., mining the shared knowledge across tasks.

5. *Lifelong Semi-Supervised Learning.* The work in this area is represented by the NELL (*Never-Ending Language Learner*) system [Carlson et al., 2010a, Mitchell et al., 2015], which has been reading the Web continuously for information extraction since January 2010, and it has accumulated millions of entities and relations.

6. *Lifelong Reinforcement Learning.* Thrun and Mitchell [1995] first proposed some LL algorithms for robot learning which tried to capture the invariant knowledge about each individual task. Tanaka and Yamamura [1997] treated each environment as a task for LL. Ring [1998] proposed a continual-learning agent that aims to gradually solve complicated tasks by learning easy tasks first. Wilson et al. [2007] proposed a hierarchical Bayesian lifelong reinforcement learning method in the framework of Markov Decision Process (MDP). Fernández and Veloso [2013] worked on policy reuse in a multi-task setting. A nonlinear feedback policy that generalizes across multiple tasks is proposed in Deisenroth et al. [2014]. Bou Ammar et al. [2014] proposed a policy gradient efficient LL algorithm following the idea in ELLA [Ruvolo and Eaton, 2013b]. This work was further enhanced with cross-domain lifelong reinforcement learning [Bou Ammar et al., 2015a] and with constraints for safe lifelong reinforcement learning [Bou Ammar et al., 2015c].

LL techniques working in other areas also exist. Silver et al. [2013] wrote an excellent survey of the early LL research published at the AAAI 2013 Spring Symposium on LL.

As we can see, although LL has been proposed for more than 20 years, research in the area has not been extensive. There could be many reasons. Some of the reasons may be as follows. First, the ML research community for the past 20 years has focused on statistical and algorithmic approaches. LL typically needs a systems approach that combines multiple components and learning algorithms. Systems approaches to learning were not in favor. This may partially explain that although the LL research has been limited, closely related paradigms of transfer learning and MTL have been researched fairly extensively because they can be done in a more statistical and algorithmic fashion. Second, much of the past ML research and applications focused on supervised learning using structured data, which are not easy for LL because there is little to be shared across tasks or domains. For example, the knowledge learned from a supervised learning system on a loan application is hard to be used in a health or education application because they do not have much in common. Also, most supervised learning algorithms generate no additional knowledge other than the final model or classifier, which is difficult to use as prior knowledge for another classification task even in a similar domain. Third, many effective ML methods such as SVM and deep learning cannot easily use prior knowledge even if such knowledge exists. These classifiers are black boxes and hard to decompose or interpret. They are generally more accurate with more training data. Fourth, related areas such as transfer learning and MTL were popular partly because they typically need only two and just a few similar tasks and datasets and do not require retention of explicit knowledge. LL, on the other hand, needs significantly more previous tasks and data in order to learn and to accumulate a large amount of explicit knowledge so that the new learning task can pick and choose the suitable knowledge to be used to help the new learning. This is analogous to human learning. If one does not have much knowledge, it is very hard for him/her to learn more knowledge. The more knowledge that one has, the easier it is for him/her to learn even more. For example, it is close to impossible for an elementary school pupil to learn graphical models. Even for an adult, if he has not studied probability theory, it is in-feasible for him to learn graphical models either.

Considering these factors, we believe that one of the more promising areas for LL is NLP due to its extensive sharing of knowledge across domains and tasks and inter-relatedness of NLP tasks as we discussed above. The text data is also abundant. Lifelong supervised, unsupervised, semi-supervised, and reinforcement learning can all be applied to text data.

1.4 DEFINITION OF LIFELONG LEARNING

The early definition of LL is as follows [Thrun, 1996b]. At any point in time, the system has learned to perform N tasks. When faced with the $(N + 1)$th task, it uses the knowledge gained from the past N tasks to help learn the $(N + 1)$th task. We extend this definition by giving it more details and additional features. First, an explicit *knowledge base* (KB) is added to retain the knowledge learned from previous tasks. Second, the ability to discover new learning tasks during model application is included. Third, the ability to learn while working (or to learn on the job) is incorporated.

Definition 1.1   *Lifelong learning* (LL) is a continuous learning process. At any point in time, the learner has performed a sequence of N learning tasks, $\mathcal{T}_1, \mathcal{T}_2, \ldots, \mathcal{T}_N$. These tasks, which are also called the *previous tasks*, have their corresponding datasets $\mathcal{D}_1, \mathcal{D}_2, \ldots, \mathcal{D}_N$. The tasks can be of different *types* and from different *domains*. When faced with the $(N + 1)$th task $\mathcal{T}_{N+1}$ (which is called the *new* or *current task*) with its data $\mathcal{D}_{N+1}$, the learner can leverage the *past knowledge* in the *knowledge base* (KB) to help learn $\mathcal{T}_{N+1}$. The task may be given or discovered by the system itself (see below). The objective of LL is usually to optimize the performance of the new task $\mathcal{T}_{N+1}$, but it can optimize any task by treating the rest of the tasks as the previous tasks. KB maintains the knowledge learned and accumulated from learning the previous tasks. After the completion of learning $\mathcal{T}_{N+1}$, KB is updated with the knowledge (e.g., intermediate as well as the final results) gained from learning $\mathcal{T}_{N+1}$. The updating can involve consistency checking, reasoning, and meta-mining of higher-level knowledge. Ideally, an LL learner should also be able to:

1. learn and function in the open environment, where it not only can apply the learned model or knowledge to solve problems but also *discover new tasks* to be learned, and

2. learn to improve the model performance in the application or testing of the learned model. This is like that after job training, we still *learn on the job* to become better at doing the job.

We can see that this definition is neither formal nor specific because LL is an emerging field and our understanding of it is still limited. For example, the research community still cannot define what knowledge is formally. We believe that it may be better to leave the definition of LL at the conceptual level rather than having it fixed or formalized. Clearly, this does not prevent us from giving a formal definition when we solve a specific LL problem. Below, we give some additional remarks.

1. The definition indicates five key characteristics of LL:

   (a) continuous learning process,

   (b) knowledge accumulation and maintenance in the KB,

   (c) the ability to use the accumulated past knowledge to help future learning,

   (d) the ability to discover new tasks, and

   (e) the ability to learn while working or to learn on the job.

   These characteristics make LL different from related learning paradigms such as transfer learning [Jiang, 2008, Pan and Yang, 2010, Taylor and Stone, 2009] and multi-task learning (MTL) [Caruana, 1997, Chen et al., 2009, Lazaric and Ghavamzadeh, 2010], which do not have one or more of these characteristics. We will discuss these related paradigms and their differences from LL in detail in Chapter 2.

   Without these capabilities, an ML system will not be able to learn in a dynamic open environment by itself, and will never be truly intelligent. By *open environment*, we mean that the application environment may contain novel objects and scenarios that have not been learned before. For example, we want to build a greeting robot for a hotel. At any point in time, the robot has learned to recognize all existing hotel guests. When it sees an existing guest, it can call him/her by his/her first name and chat. It must also detect any new guests that it has not seen before. On seeing a new guest, it can say hello, ask for his/her name, take many pictures, and learn to recognize the guest. Next time when it sees the new guest again, it can call him/her by his/her first name and chat like an old friend. The real-world road environment for self-driving cars is another very typical dynamic and open environment.

2. Since knowledge is accumulated and used in LL, this definition forces us to think about the issue of prior knowledge and the role it plays in learning. LL thus brings in many other aspects of Artificial Intelligence to ML, e.g., knowledge representation, acquisition, reasoning, and maintenance. Knowledge, in fact, plays a central rule. It not only can help improve future learning, but can also help collect and label training data (self-supervision) and discover new tasks to be learned in order to achieve autonomy in learning. The integration of both data-driven learning and knowledge-driven learning is probably what human learning is all about. The current ML focuses almost entirely on data-driven optimization

learning, which we humans are not good at. Instead, we are very good at learning based on our prior knowledge. It is well-known that for human beings, the more we know the more we can learn and the easier we can learn. If we do not know anything, it is very hard to learn anything. ML research should thus pay more attention to knowledge and build machines that learn like humans.[1]

3. We distinguish two types of tasks.

   (a) *Independent tasks*: Each task $\mathcal{T}_i$ is independent of the other tasks. This means that each task can be learned independently, although due to their similarities and sharing of some latent structures or knowledge, learning $\mathcal{T}_i$ can leverage the knowledge gained from learning previous tasks.

   (b) *Dependent tasks*: Each task $\mathcal{T}_i$ has some dependence on some other tasks. For example, in open-world learning (Chapter 5)[Fei et al., 2016], each new supervised learning task adds a new class to the previous classification problem, and needs to build a new multi-class classifier that is able to classify data from all previous and the current classes.

4. The tasks do not have to be from the same domain. Note that there is still no unified definition of a *domain* in the literature that is applicable to all areas. In most cases, the term is used informally to mean a setting with a fixed feature space where there can be multiple different tasks of the same type or of different types (e.g., information extraction, coreference resolution, and entity linking). Some researchers even use domain and task interchangeably because there is only one task from each domain in their study. We also use them interchangeably in many cases in this book due to the same reason but will distinguish them when needed.

5. The shift to the new task can happen abruptly or gradually, and the tasks and their data do not have to be provided by some external systems or human users. Ideally, a lifelong learner should also be able to find its own learning tasks and training data in its interaction with humans and the environment or using its previously learned knowledge to perform open-world and self-supervised learning.

6. The definition indicates that LL may require a *systems approach* that combines multiple learning algorithms and different knowledge representation schemes. It is unlikely that a single learning algorithm is able to achieve the goal of LL. LL, in fact, represents a large and rich problem space. Much research is needed to design algorithms to achieve each capability or characteristic.

---

[1]If a learning system can do both data-driven optimization and human-level knowledge-based learning, we may say that it has achieved some kind of *super learning* capability, which may also mean that it has reached some level of *artificial general intelligence* (AGI) because we humans certainly cannot do learning based on large scale data-driven optimization or remember a large quantity of knowledge in our brain as a machine can.

Based on Definition 1.1, we can outline a general process of LL and an LL system architecture, which is very different from that of the current *isolated learning* paradigm with only a single task T and dataset D. Figure 1.1 illustrates the classic isolated learning paradigm, where the learned model is used in its intended application.

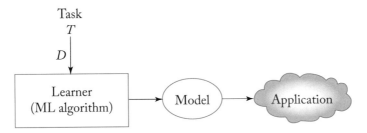

Figure 1.1: The classic machine learning paradigm.

The new LL system architecture is given in Figure 1.2. Below, we first describe the key components of the system and then the LL process. We note that this general architecture is for illustration purposes. Not all existing systems use all the components or sub-components. In fact, most current systems are much simpler. Moreover, there is still not a generic LL system that can perform LL in all possible domains for all possible types of tasks. In fact, we are still far from that. Unlike many ML algorithms such as SVM and deep learning, which can be applied to any learning task as long as the data is represented in a specific format required by these algorithms, current LL algorithms are still specific to some types of tasks and data.

1. **Knowledge Base (KB)**: It is mainly for storing the previously learned knowledge. It has a few sub-components.

   (a) *Past Information Store* (PIS): It stores the information resulted from the past learning, including the resulting models, patterns, or other forms of outcome. PIS may involve sub-stores for information such as (1) the original data used in each previous task, (2) intermediate results from each previous task, and (3) the final model or patterns learned from each previous task. As for what information or knowledge should be retained, it depends on the learning task and the learning algorithm. For a particular system, the user needs to decide what to retain in order to help future learning.

   (b) *Meta-Knowledge Miner* (MKM). It performs meta-mining of the knowledge in the PIS and in the meta-knowledge store (see below). We call this *meta-mining* because it mines higher-level knowledge from the saved knowledge. The resulting knowledge is stored in the Meta-Knowledge Store. Here multiple mining algorithms may be used to produce different types of results.

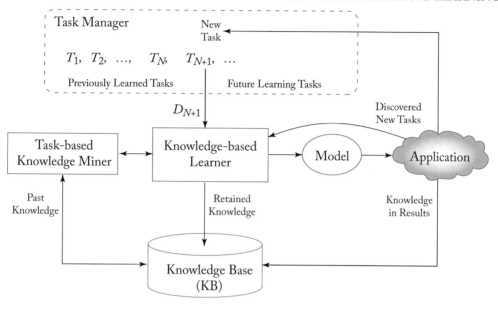

Figure 1.2: The lifelong machine learning system architecture.

(c) *Meta-Knowledge Store* (MKS): It stores the knowledge mined or consolidated from PIS and also from MKS itself. Some suitable knowledge representation schemes are needed for each application.

(d) *Knowledge Reasoner* (KR): It makes inference based on the knowledge in MKB and PIS to generate more knowledge. Most current systems do not have this sub-component. However, with the advance of LL, this component will become increasingly important.

Since the current LL research is still in its infancy, as indicated above, none of the existing systems has all these sub-components.

2. **Knowledge-Based Learner (KBL)**: For LL, it is necessary for the learner to be able to use prior knowledge in learning. We call such a learner a *knowledge-based learner*, which can leverage the knowledge in the KB to learn the new task. This component may have two sub-components: (1) *Task knowledge miner* (TKM), which makes use of the raw knowledge or information in the KB to mine or identify knowledge that is appropriate for the current task. This is needed because in some cases, KBL cannot use the raw knowledge in the KB directly but needs some task-specific and more general knowledge mined from the KB [Chen and Liu, 2014a,b], and (2) the *learner* that can make use of the mined knowledge in learning.

3. **Task-based Knowledge Miner (TKM)**: This module mines knowledge from the KB specifically for the new task.

4. **Model**: This is the learned model, which can be a prediction model or classifier in supervised learning, clusters or topics in unsupervised learning, a policy in reinforcement learning, etc.

5. **Application**: This is the real-world application for the model. It is important to note that during model application, the system can still learn new knowledge (i.e., "knowledge in results"), and possibly discover new tasks to be learned. Application can also give feedback to the knowledge-based learner for model improvement.

6. **Task Manager (TM)**: It receives and manages the tasks that arrive in the system, handles the task shift, and presents the new learning task to the KBL in a lifelong manner.

Lifelong Learning Process: A typical LL process starts with the Task Manager assigning a new task to the KBL (the task can be given or discovered automatically). KBL then works with the help of the past knowledge stored in the KB to produce an output model for the user and also send the information or knowledge that needs to be retained for future use to the KB. In the application process, the system may also discover new tasks and learn while working (learn on the job). Some knowledge gained in applications can also be retained to help future learning.

1.5   TYPES OF KNOWLEDGE AND KEY CHALLENGES

Definition 1.1 does not give any detail about what knowledge or its representation form is in the KB. This is mainly due to our limited understanding. There is still no well-accepted definition of knowledge or its general representation scheme. In the current LL research, past knowledge usually serves as some kind of prior information (e.g., prior model parameters or prior probabilities) for the new task. Each existing paper uses one or two specific forms of knowledge that are suitable for its proposed techniques and intended applications. For example, some methods use a set of shared latent parameters [Ruvolo and Eaton, 2013b, Wilson et al., 2007] as knowledge. Some directly use model parameters of previous tasks as knowledge [Chen et al., 2015, Shu et al., 2016]. Some use previous model application results as knowledge, e.g., topics from topic modeling [Chen and Liu, 2014a, Chen et al., 2015] and items extracted from previous information extraction models [Liu et al., 2016, Shu et al., 2017b]. Some even use past relevant data as knowledge to augment the new task data [Xu et al., 2018]. Knowledge is usually represented based on how it is used in individual algorithms. There are still no general knowledge representation schemes that can be used across different algorithms or different types of tasks. Definition 1.1 also does not specify how to maintain or update the KB. For a particular LL algorithm and a particular form of shared knowledge, one needs to design a KB and its maintenance or updating methods based on the algorithm and its knowledge representation need.

There are mainly two types of shared knowledge that are used in learning the new task.

1. *Global knowledge*: Many existing LL methods assume that there is a *global latent structure* among tasks that is shared by all tasks [Bou Ammar et al., 2014, Ruvolo and Eaton, 2013b, Thrun, 1996b, Wilson et al., 2007] (Sections 3.2, 3.4, 9.1, 9.2, and 9.3). This global structure can be learned and leveraged in the new task learning. The approaches based on global knowledge transfer and sharing mainly grew out of or inspired by MTL, which jointly optimizes the learning of multiple similar tasks. Such knowledge is more suitable for similar tasks in the same domain because such tasks are often highly correlated or have very similar distributions.

2. *Local knowledge*: Many other methods do not assume such a global latent structure among tasks [Chen and Liu, 2014a,b, Chen et al., 2015, Fei et al., 2016, Liu et al., 2016, Shu et al., 2016, Tanaka and Yamamura, 1997] (Sections 3.5, 5.2, 6.2, 6.3, 7.1, 7.2, 7.3, and 7.4). Instead, during the learning of a new task they pick and choose the pieces of knowledge learned from previous tasks to use based on the need of the current task. This means that different tasks may use different pieces of knowledge learned from different previous tasks. We call such pieces of knowledge the *local knowledge* because they are local to their individual previous tasks and are not assumed to form a coherent global structure. Local knowledge is likely to be more suitable for related tasks from different domains because the shared knowledge from any two domains may be small. But the prior knowledge that can be leveraged by the new task can still be large because the prior knowledge can be from many past domains.

LL methods based on local knowledge usually focus on optimizing the current task performance with the help of past knowledge. They can also be used to improve the performance of any previous task by treating that task as the new/current task. The main advantage of these methods is their flexibility as they can choose whatever pieces of past knowledge that are useful to the new task. If nothing is useful, the past knowledge will not be used. The main advantage of LL methods based on global knowledge is that they often approximate optimality on all tasks, including the previous and the current ones. This property is inherited from MTL. However, when the tasks are highly diverse and/or numerous, this can be difficult.

As the previous learned knowledge is involved, apart from the classic issues about knowledge discussed above (e.g., what knowledge to retain, how to represent and use the knowledge, and how to maintain the KB), there are two other fundamental challenges about knowledge in LL. We will describe how some existing techniques deal with these challenges later in the book.

1. *Correctness of knowledge*: Clearly, using incorrect past knowledge is detrimental to the new task learning. In a nutshell, LL can be regarded as a continuous bootstrapping process. Errors can propagate from previous tasks to subsequent tasks to generate more and more errors. We humans seem to have a good idea of what is correct or what is incorrect. But there is still no satisfactory technique for detecting wrong knowledge. Many existing papers do not deal with this challenge [Silver and Mercer, 2002, Silver et al., 2015, Thrun,

1996b] or deal with it implicitly to some extent [Ruvolo and Eaton, 2013b, Wilson et al., 2007]. There are also many papers that deal with the challenge explicitly [Chen and Liu, 2014a,b, Chen et al., 2015, Liu et al., 2016, Mitchell et al., 2015, Shu et al., 2016]. For example, one strategy is to find those pieces of knowledge that have been discovered in many previous tasks/domains [Chen and Liu, 2014a,b, Chen et al., 2015, Shu et al., 2016]. Another strategy is to make sure that the piece of knowledge is discovered from different contexts using different techniques [Mitchell et al., 2015]. Although these and other strategies are useful, they are still not satisfactory because of two main issues. First, they are not foolproof because they can still produce wrong knowledge. Second, they have low recall because most pieces of correct knowledge cannot pass these strategies and thus cannot be used subsequently, which prevents LL from producing even better results. We will detail these strategies when we discuss the related papers.

2. *Applicability of knowledge.* Although a piece of knowledge may be correct in the context of some previous tasks, it may not be applicable to the current task. Application of inappropriate knowledge has the same negative consequence as the above case. Again, we humans are quite good at recognizing the right context for the application of a piece of knowledge, which is very difficult for automated systems. Again, many papers do not deal with the challenge, however, some do, e.g., Chen and Liu [2014a], Chen et al. [2015], Shu et al. [2016], and Xu et al. [2018]. We will describe them when we discuss these papers as they are quite involved.

Clearly, the two challenges are closely related. It is seemingly that we only need to be concerned with the applicability challenge regardless whether the knowledge is correct or not because if a piece of knowledge is not correct, it cannot be applicable to the new task. This is often not the case because in deciding applicability, we may just be able to decide whether a new task or domain context is similar to some older tasks or domain contexts. If so, we can use the knowledge gained from those older tasks. Then we must make sure that the knowledge from those older tasks is correct.

1.6   EVALUATION METHODOLOGY AND ROLE OF BIG DATA

Unlike the classic isolated learning where the evaluation of a learning algorithm is based on training and testing using data from the same task/domain, LL needs a different evaluation methodology because it involves a sequence of tasks and we want to see improvements in the learning of new tasks. Experimental evaluation of an LL algorithm in the current research is commonly done using the following steps.

1. *Run on the data from the previous tasks*: We first run the algorithm on the data from a set of previous tasks, one at a time in a given sequence, and retain the knowledge gained

in the KB. Obviously, there can be multiple variations or versions of the algorithm (e.g., with different types of knowledge used and more or less knowledge used) that can be experimented with.

2. *Run on the data of the new task*: We then run the LL algorithm on the new task data by leveraging the knowledge in the KB.

3. *Run baseline algorithms*: For comparison, we run some baseline algorithms. There are usually two kinds of baselines. The first kind are algorithms that perform isolated learning on the new data without using any past knowledge. The second kind are existing LL algorithms.

4. *Analyze the results*: This step compares the results from steps 2 and 3 and analyzes the results to make some observations, e.g., to show that the results from the LL algorithm in step 2 are superior to those from the baselines in step 3.

There are several additional considerations in carrying out an LL experimental evaluation.

1. *A large number of tasks*: A large number of tasks and datasets are needed to evaluate an LL algorithm. This is because the knowledge gained from a few tasks may not be able to improve the learning of the new task much as each task may only provide a very small amount of knowledge that is useful to the new task (unless all the tasks are very similar) and the data in the new task is often quite small.

2. *Task sequence*: The sequence of the tasks to be learned can be significant, meaning that different task sequences can generate different results. This is so because LL algorithms typically do not guarantee optimal solutions for all previous tasks. To take the sequence effect into consideration in the experiment, one can try several random sequences of tasks and generate results for the sequences. The results can then be aggregated for comparison purposes. Existing papers mainly use only one random sequence in their experiments.

3. *Progressive experiments*: Since more previous tasks generate more knowledge, and more knowledge in turn enables an LL algorithm to produce better results for the new task, it is thus desirable to show how the algorithm performs on the new task as the number of previous tasks increases.

Note that it is not our intention to cover all possible kinds of evaluations in the current research on LL. Our purpose is simply to introduce the common evaluation methodologies. In evaluating a specific algorithm, one has to consider the special characteristics of the algorithm (such as its assumptions and parameter settings) and the related research in order to design a comprehensive set of experiments.

Role of Big Data in LL Evaluation: It is common knowledge that the more you know the more you can learn and the easier you can learn. If we do not know anything, it is very hard to learn anything. These are intuitive as each one of us must have experienced this in our lives. The same is true for a computer algorithm. Thus, it is important for an LL system to learn from a diverse range and a large number of domains to give the system a wide vocabulary and a wide range of knowledge so that it can help learn in diverse future domains. Furthermore, unlike transfer learning, LL needs to automatically identify the pieces of past knowledge that it can use, and not every past task/domain is useful to the current task. LL experiments and evaluation thus require data from a large number of domains or tasks and consequently large volumes of data. Fortunately, big datasets are now readily available in many applications such as image and text that can be used in LL evaluations.

1.7   OUTLINE OF THE BOOK

This book introduces and surveys this important and emerging field. Although the body of literature is not particularly large, related papers are published in a large number of conferences and journals. There is also a large number of papers that do not exhibit all the characteristics of LL, but are related to it to some extent. It is thus hard, if not impossible, to cover all of the important work in the field. As a result, this book should not be taken to be an exhaustive account of everything in the field.

The book is organized as follows. In Chapter 2, we discuss some related ML paradigms to set the stage and background. We will see that these existing paradigms are different from LL because they lack one or more of the key characteristics of LL. However, all these paradigms involve some forms of knowledge sharing or transfer across tasks and can even be made continual in some cases. Thus, we regard LL as an advanced ML paradigm that extends these existing paradigms in the progression of making ML more intelligent and closer to human learning.

In Chapter 3, we focus on discussing existing research on supervised LL, where we will see a fairly detailed account of some early and more recent supervised LL methods. In Chapter 4, we continue the discussion of supervised LL but in the context of deep neural networks (DNNs), where LL is sometimes also called *continual learning*. The main goal in this context is to solve the *catastrophic forgetting* problem in deep learning when learning multiple tasks. Chapter 5 is another chapter related to supervised learning. However, as its name suggests, this type of learning learns in the open-world, where the test data may contain instances from unseen classes (not seen in training). This is in contrast to the classic closed-world learning, where all test classes have appeared in training.

In Chapter 6, we discuss lifelong topic models. In these models, discovered topics from previous tasks are mined to extract reliable knowledge that can be exploited in the new model inferencing to generate better topics for the new task. Chapter 7 discusses lifelong information extraction. Information extraction is a very suitable problem for LL because information extracted in the past is often quite useful for future extraction due to knowledge sharing across tasks and/or domains. Chapter 8 switches topic and discusses a preliminary work about lifelong interactive knowledge learning in human-machine conversation. This is a new direction as existing chatbots cannot learn new knowledge after they are built or deployed. Lifelong reinforcement learning is covered in Chapter 9. Chapter 10 concludes the book and discusses some major challenges and future directions of the LL research.

CHAPTER 2

Related Learning Paradigms

As described in the introduction chapter, lifelong learning (LL) has several key characteristics: continuous learning process, explicit knowledge retention and accumulation, and the use of the previously learned knowledge to help learn new tasks. Additionally, it is also desirable to have the ability to discover new tasks and learn them incrementally, and to learn additional knowledge during the real-life applications to improve the model. There are several machine learning (ML) paradigms that have related characteristics. This chapter discusses the most related ones, i.e., transfer learning or domain adaption, multi-task learning (MTL), online learning, reinforcement learning, and meta-learning. The first two paradigms are more closely related to LL because they both involve knowledge transfer across domains or tasks, but they don't learn continuously and don't retain or accumulate learned knowledge explicitly. Online learning and reinforcement learning involves continuous learning processes but they still focus on a single learning task with a time dimension. Meta-learning is also concerned with multiple tasks, with primary focus on one-shot or few-shot learning. These differences will become clearer after we review some representative techniques for each of these related learning paradigms.

2.1   TRANSFER LEARNING

Transfer learning is a popular topic of research in ML and data mining. It is also commonly known as *domain adaptation* in natural language processing. It usually involves two domains: a *source domain* and a *target domain*. Although there can be more than one source domain, in almost all existing research only one source domain is used. The source domain normally has a large amount of labeled training data while the target domain has little or no labeled training data. The goal of transfer learning is to use the labeled data in the source domain to help learning in the target domain (see three excellent surveys of the area [Jiang, 2008, Pan and Yang, 2010, Taylor and Stone, 2009]). Note that in the literature, some researchers also use the terms *source task* and *target task* rather than *source domain* and *target domain*, but by far, the latter terminologies are more commonly used as the source, and the target tasks are often from different domains or quite different distributions [Pan and Yang, 2010].

There are many types of knowledge that can be transferred from the source domain to the target domain to help learning in the target domain. For example, Bickel et al. [2007], Dai et al. [2007b,c], Jiang and Zhai [2007], Liao et al. [2005], and Sugiyama et al. [2008] directly treated certain parts of data instances in the source domain as the knowledge with instance reweighing and importance sampling and transfer it over to the target domain. Ando and Zhang

[2005], Blitzer et al. [2006, 2007], Dai et al. [2007a], Daume III [2007], and Wang and Mahadevan [2008] used features from the source domain to generate new feature representations for the target domain. Bonilla et al. [2008], Gao et al. [2008], Lawrence and Platt [2004], and Schwaighofer et al. [2004] transferred learning parameters from the source domain to the target domain. To give a flavor of transfer learning, we briefly discuss some existing transfer learning methods below.

2.1.1   STRUCTURAL CORRESPONDENCE LEARNING

One of the popular transfer learning techniques is the Structural Correspondence Learning (SCL) proposed in Blitzer et al. [2006, 2007]. This method is mainly used in text classification. The algorithm works as follows: given labeled data from the source domain and unlabeled data from both the source and target domains, SCL tries to find a set of *pivot* features that have the same characteristics or behaviors in both domains. If a non-pivot feature is correlated with many of the same pivot features across different domains, this feature is likely to behave similarly across different domains. For example, if a word w co-occurs very frequently with the same set of *pivot* words in both domains, then w is likely to behave the same (e.g., holding the same semantic meaning) across domains.

To implement the above idea, SCL first chooses a set of m features which occur frequently in both domains and are also good predictors of the source label (in their paper these were the features with the highest mutual information with the source label). These pivot features represent the shared feature space of the two domains. SCL then computes the correlations of each pivot feature with other non-pivot features in both domains. This produces a correlation matrix $\mathbf{W}$ where row i is a vector of correlation values of non-pivot features with the ith pivot feature. Intuitively, positive values indicate that those non-pivot features are positively correlated with the ith pivot feature in the source domain or in the target domain. This establishes a feature correspondence between the two domains. After that, singular value decomposition (SVD) is employed to compute a low-dimensional linear approximation θ (the top k left singular vectors, transposed) of $\mathbf{W}$. The final set of features for training and for testing is the original set of features $\mathbf{x}$ combined with $\theta\mathbf{x}$ which produces k real-valued features. The classifier built using the combined features and the labeled data in the source domain should work in both the source and the target domains.

Pan et al. [2010] proposed a method similar to SCL at the high level. The algorithm works in the setting where there are only labeled examples in the source domain and unlabeled examples in the target domain. It bridges the gap between the domains by using a spectral feature alignment (SFA) algorithm to align domain-specific words from different domains into some unified clusters, with domain-independent words as the bridge. Domain-independent words are like pivot words above and can be selected similarly.

2.1.2   NAÏVE BAYES TRANSFER CLASSIFIER

Many transfer learning methods have been proposed in the context of Naïve Bayesian (NB) classification [Chen et al., 2013a, Dai et al., 2007b, Do and Ng, 2005, Rigutini et al., 2005]. Here we briefly describe the work in Dai et al. [2007b] to give a flavor of such methods.

Dai et al. [2007b] proposed a method called *Naïve Bayes Transfer Classifier* (NBTC). Let the labeled data from the source domain be $\mathcal{D}_l$ with the distribution $\mathfrak{D}_l$, and the unlabeled data from the target domain be $\mathcal{D}_u$ with the distribution $\mathfrak{D}_u$. $\mathfrak{D}_l$ may not be the same as $\mathfrak{D}_u$. A two-step approach is employed in NBTC.

1. Build an initial Naïve Bayesian classifier using the labeled data $\mathcal{D}_l$ under $\mathfrak{D}_l$ from the source domain.

2. Run an Expectation-Maximization (EM) algorithm together with the target unlabeled data to find a local optimal model under the target domain distribution $\mathfrak{D}_u$.

The objective function of NBTC is as follows, which aims to find a local optimum of the *Maximum a Posteriori* (MAP) hypothesis under $\mathfrak{D}_u$:

$$h_{MAP} = \underset{h}{\operatorname{argmax}}\, P_{\mathfrak{D}_u}(h) \times P_{\mathfrak{D}_u}(\mathcal{D}_l, \mathcal{D}_u | h) \ . \tag{2.1}$$

This equation considers the probability of the source domain labeled data and the target domain unlabeled data under the hypothesis h. The labeled data provides the supervised information, while estimating the probability of the unlabeled data under $\mathfrak{D}_u$ ensures that the model fits for $\mathcal{D}_u$. Based on Bayes' rule, NBTC maximizes the log-likelihood $l(h|\mathcal{D}_l, \mathcal{D}_u) = \log P_{\mathfrak{D}_u}(h|\mathcal{D}_l, \mathcal{D}_u)$,

$$\begin{aligned}
l(h|\mathcal{D}_l, \mathcal{D}_u) \propto\ & \log P_{\mathfrak{D}_u}(h) \\
& + \sum_{d \in \mathcal{D}_l} \log \sum_{c \in C} P_{\mathfrak{D}_u}(d|c, h) \times P_{\mathfrak{D}_u}(c|h) \\
& + \sum_{d \in \mathcal{D}_u} \log \sum_{c \in C} P_{\mathfrak{D}_u}(d|c, h) \times P_{\mathfrak{D}_u}(c|h) \ ,
\end{aligned} \tag{2.2}$$

where C is the set of classes and $d \in \mathcal{D}_l$ is a document in $\mathcal{D}_l$. To optimize it, Dai et al. [2007b] applied the EM algorithm as follows:

- **E-Step**:

$$P_{\mathfrak{D}_u}(c|d) \propto P_{\mathfrak{D}_u}(c) \prod_{w \in d} P_{\mathfrak{D}_u}(w|c) \tag{2.3}$$

- **M-Step**:

$$P_{\mathfrak{D}_u}(c) \propto \sum_{i \in \{l, u\}} P_{\mathfrak{D}_u}(\mathcal{D}_i) \times P_{\mathfrak{D}_u}(c|\mathcal{D}_i) \tag{2.4}$$

$$P_{\mathfrak{D}_u}(w|c) \propto \sum_{i \in \{l,u\}} P_{\mathfrak{D}_u}(\mathcal{D}_i) \times P_{\mathfrak{D}_u}(c|\mathcal{D}_i) \times P_{\mathfrak{D}_u}(w|c, \mathcal{D}_i) \ , \tag{2.5}$$

where $w \in d$ represents a word in document d. $P_{\mathfrak{D}_u}(c|\mathcal{D}_i)$ and $P_{\mathfrak{D}_u}(w|c, \mathcal{D}_i)$ can be rewritten via the NB classification formulation (see Dai et al. [2007b] for more details). The above E-step and M-step are repeated to reach a local optimal solution.

Chen et al. [2013a] proposed two EM-type algorithms called FS-EM (Feature Selection EM) and Co-Class (Co-Classification). FS-EM uses feature selection as the mechanism to transfer knowledge from the source domain to the target domain in each EM iteration. Co-Class further adds the idea of co-training [Blum and Mitchell, 1998] to deal with the imbalance of the shared positive and negative features. It builds two NB classifiers, one on labeled data, and the other on the unlabeled data with predicted labels. An earlier work for cross-language text classification also used a similar idea in the context of NB classification [Rigutini et al., 2005], which transfers knowledge from the labeled data in English to the unlabeled data in Italian.

2.1.3 DEEP LEARNING IN TRANSFER LEARNING

In recent years, deep learning or deep neural network (DNN) has emerged as a major learning method and has achieved very promising results [Bengio, 2009]. It has been used by several researchers for transfer learning.

For example, instead of using the traditional raw input as features which may not generalize well across domains, Glorot et al. [2011] proposed to use the low-dimensional features learned using deep learning to help prediction in the new domain. In particular, Stacked Denoising Auto-encoder of Vincent et al. [2008] was employed in Glorot et al. [2011]. In an auto-encoder, there are typically two functions: an encoder function $h()$ and a decoder function $g()$. The reconstruction of input x is given by $r(x) = g(h(x))$. To train an auto-encoder, the objective function is to minimize the reconstruction error $loss(x, r(x))$. Then, auto-encoders can be trained and stacked together as a hierarchical network. In this network, the auto-encoder at level i takes the output of the $(i-1)th$ auto-encoder as input. Level 0 takes the raw input. In denoising an auto-encoder, the input vector x is stochastically corrupted into another vector $\hat{x}$ and the objective function is to minimize a denoising reconstruction error loss $loss(x, r(\hat{x}))$. In Glorot et al. [2011], the model is learned in a greedy layer-wise fashion using stochastic gradient descent. The first layer uses logistic sigmoid to transform the raw input. For the upper layers, the softplus activation function, $\log(1 + \exp(x))$, is used. After learning the auto-encoders, a linear SVM with squared hinge loss is trained on the labeled data from the source domain and tested on the target domain.

Yosinski et al. [2014] studied the transferability of features in each layer of a DNN. They argued that the lowest level or the raw input layer is very *general* as it is independent of the task and the network. In contrast, the features from the highest level depend on the task and cost function, and thus are *specific*. For example, in a supervised learning task, each output unit corresponds to a particular class. From the lowest level to the highest level, there is a transfer from generality to specificity. To experiment the transferability of features in each layer in a DNN, they trained a neural network from the source domain and copied the first n layers to the neural network for the target domain. The remaining layers in the target neural network are randomly initialized. They showed that transferred features in the neural network from the source domain are indeed helpful to the target domain learning. Also in the transfer learning setting, Bengio [2012] focused on unsupervised pre-training of representations and discussed potential challenges of deep learning for transfer learning.

2.1.4 DIFFERENCE FROM LIFELONG LEARNING

Transfer learning is different from LL in the following aspects. We note that since the literature on transfer learning is extensive, the differences described here may not be applicable to every individual transfer learning paper.

1. Transfer learning is not concerned with continuous learning or knowledge accumulation. Its transfer of information or knowledge from the source domain to the target domain is usually one time only. It does not retain the transferred knowledge or information for future use. LL, on the other hand, represents continuous learning, and knowledge retention and accumulation are essential for LL as they not only enable the system to become more and more knowledgeable, but also allow it to learn additional knowledge more accurately and easily in the future.

2. Transfer learning is unidirectional. It transfers knowledge from only the source domain to the target domain, but not the other way around because the target domain has little or no training data. In LL, the learning result from the new domain or task can be used to improve learning in previous domains or tasks if needed.

3. Transfer learning typically involves only two domains, a source domain and a target domain (although in some cases there is more than one source domain). It assumes that the source domain is very similar to the target domain; otherwise the transfer can be detrimental. The two similar domains are usually selected by human users. LL, on the other hand, considers a large (possibly unlimited) number of tasks/domains. In solving a new problem, the learner needs to decide what past knowledge is appropriate for the new learning task. It does not have the assumption made by transfer learning. In LL, if there is useful knowledge from the past, use it. If not, just learn using the current domain data. Since LL typically involves a large number of past domains or tasks, the system can accumulate a large amount of past knowledge such that the new learning task is very likely to find some pieces of past knowledge useful.

4. Transfer learning does not identify new tasks to be learned during model application (after the model has been built) [Fei et al., 2016], or learn on the job, i.e., learn while working or model application [Shu et al., 2017b].

2.2   MULTI-TASK LEARNING

Multi-task learning (MTL) learns multiple related tasks simultaneously, aiming at achieving a better performance by using the relevant information shared by multiple tasks [Caruana, 1997, Chen et al., 2009, Li et al., 2009]. The rationale is to introduce inductive bias in the joint hypothesis space of all tasks by exploiting the task relatedness structure. It also prevents overfitting in the individual task and thus has a better generalization ability. Note that unlike in transfer learning, we mostly use the term *multiple tasks* rather than *multiple domains* as much of the existing research in the area is based on multiple similar tasks from the same domain of application. We now define *multi-task learning*, which is also referred to as *batch multi-task learning*.

Definition 2.1   *Multi-task Learning* (MTL) is concerned with learning multiple tasks $\mathcal{T} = \{1, 2, \ldots, N\}$ simultaneously. Each task $t \in \mathcal{T}$ has its training data $\mathcal{D}^t$. The goal is to maximize the performance across *all* tasks.

Since most existing works on MTL focused on supervised learning, here we discuss only multi-task supervised learning. Let each task t have the training data $\mathcal{D}^t = \{(\mathbf{x}_i^t, y_i^t) : i = 1, \ldots, n_t\}$, where n_t is the number of training instances in $\mathcal{D}^t$. $\mathcal{D}^t$ is defined by a hidden (or latent) true mapping $\hat{f}^t(\mathbf{x})$ from an instance space $\mathcal{X}^t \subseteq \mathbb{R}^d$ to a set of labels $\mathcal{Y}^t$ ($y_i^t \in \mathcal{Y}^t$) (or $\mathcal{Y}^t = \mathbb{R}$ for regression). d is the feature dimension. We want to learn a mapping function $f^t(\mathbf{x})$ for each task t so that $f^t(\mathbf{x}) \approx \hat{f}^t(\mathbf{x})$. Formally, given a loss function $\mathcal{L}$, multi-task learning minimizes the following objective function:

$$\sum_{t=1}^{N} \sum_{i=1}^{n_t} \mathcal{L}\left(f(\mathbf{x}_i^t), \ y_i^t\right) \ . \tag{2.6}$$

In contrast to this batch MTL, *online multi-task learning* aims to learn the tasks sequentially and accumulate knowledge over time and leverage the knowledge to help subsequent learning (or to improve some previous learning task). Online MTL is thus LL.

2.2.1   TASK RELATEDNESS IN MULTI-TASK LEARNING

MTL assumes that tasks are closely *related*. There are different assumptions in terms of *task relatedness*, which lead to different modeling approaches.

Evgeniou and Pontil [2004] assumed that all data for the tasks come from the same space and all the task models are close to a global model. Under this assumption, they modeled the relation between tasks using a task-coupling parameter with regularization. Baxter [2000] and Ben-David and Schuller [2003] assumed that the tasks share a common underlying representation, e.g., using a common set of learned features. Some other works used probabilistic approaches assuming that the parameters share a common prior [Daumé III, 2009, Lee et al., 2007, Yu et al., 2005].

Task parameters can also lie in a low-dimensional subspace, which is shared across tasks [Argyriou et al., 2008]. Instead of assuming all tasks sharing the full space, Argyriou et al. [2008] assumed that they share a low rank of the original space. However, the low rank assumption does not distinguish tasks. When some unrelated tasks are considered, the performance can deteriorate. To address this issue, some papers assumed that there are disjoint groups of tasks and applied clustering to group tasks [Jacob et al., 2009, Xue et al., 2007]. The tasks within a cluster are considered similar to each other. On the other hand, Yu et al. [2007] and Chen et al. [2011] assumed that there is a group of related tasks while the unrelated tasks are a small number of outliers. Gong et al. [2012] assumed that the related tasks share a common set of features while the outlier tasks do not. Kang et al. [2011] incorporated grouping structures using regularization. However, only the tasks in the same group are modeled together, so the possible sharing structure between tasks from different groups is ignored.

Recently, Kumar et al. [2012] assumed that the parameter vector of each task is a linear combination of a finite number of underlying basis or latent components. Instead of using the assumption of disjoint task groups [Jacob et al., 2009, Xue et al., 2007], they assumed that the tasks in different groups can overlap with each other in one or more bases. Based on this idea, they proposed a MTL model called GO-MTL. We detail it in the next subsection. Maurer et al. [2013] proposed to use sparse coding and dictionary learning in multi-task learning. Extending GO-MTL, Ruvolo and Eaton [2013b] proposed the Efficient Lifelong Learning Algorithm (ELLA) that dramatically improves the efficiency and makes it an online MTL method, which is regarded as an LL method as it satisfies the LL definition. We will introduce ELLA in Section 3.4.

2.2.2   GO-MTL: MULTI-TASK LEARNING USING LATENT BASIS

Grouping and Overlap in Multi-Task Learning (GO-MTL) [Kumar et al., 2012] takes a parametric approach to model building in which the model or the prediction function $f^t(\mathbf{x}) = f^t(\mathbf{x}; \boldsymbol{\theta}^t)$ for each task t is governed by the task-specific parameter vector $\boldsymbol{\theta}^t \in \mathbb{R}^d$, where d is the dimension of the data. Given N tasks, GO-MTL assumes that there are k $(< N)$ *latent basis model components* among the models of the multiple tasks. Each basis model component $\mathbf{L}$ is represented by a vector of size d. The k basis model components are represented by a $d \times k$ matrix $\mathbf{L} = (\mathbf{L}_1, \dots, \mathbf{L}_k)$. The parameter vector $\boldsymbol{\theta}^t$ of the model for each task t is assumed to be a linear combination of the k basis model components and the weight vector $\mathbf{s}^t$, i.e., $\boldsymbol{\theta}^t = \mathbf{L}\mathbf{s}^t$, and $\mathbf{s}^t$ is assumed to be sparse. Considering all the tasks, we have:

$$\underset{d \times N}{\boldsymbol{\Theta}} = \underset{d \times k}{\mathbf{L}} \times \underset{k \times N}{\mathbf{S}} , \qquad (2.7)$$

where $\boldsymbol{\Theta} = [\boldsymbol{\theta}^1, \boldsymbol{\theta}^2, \dots, \boldsymbol{\theta}^N]$ and $\mathbf{S} = [\mathbf{s}^1, \mathbf{s}^2, \dots, \mathbf{s}^N]$.

The idea is that each task can be represented by *some* of the basis model components. This mechanism takes into consideration both related and unrelated tasks. A pair of related tasks will lead to the overlapping of their linear weight vectors, while two unrelated tasks can be distinguished via their little linear weight vector overlapping. Thus, GO-MTL does not assume disjointed groups of tasks like Jacob et al. [2009] and Xue et al. [2007]. As discussed above, the disadvantage of disjoint groups is that the tasks from different groups will not have interactions with each other. However, it is possible that although the tasks are in different groups, they may be negatively correlated or they still share some information, both of which can be useful for MTL. The partial overlap among tasks is thus allowed in GO-MTL, which is flexible in dealing with sophisticated task relatedness without strong assumptions.

Objective Function

Given the training data $\mathcal{D}^t$ for each task t, the objective function is to minimize the predictive loss over all tasks while encouraging the sharing of structures between the tasks, which is defined as follows:

$$\sum_{t=1}^{N}\sum_{i=1}^{n_t} \mathcal{L}\left(f(\mathbf{x}_i^t; \mathbf{Ls}^t), y_i^t\right) + \mu \|\mathbf{S}\|_1 + \lambda \|\mathbf{L}\|_F^2 \quad , \tag{2.8}$$

where $\mathcal{L}$ is the empirical loss function, $(\mathbf{x}_i^t, y_i^t)$ is the ith labeled instance in the training data for task t. The function f is $f(\mathbf{x}_i^t; \mathbf{Ls}^t) = \boldsymbol{\theta}^t \mathbf{x}_i^t = (\mathbf{Ls}^t)^{\mathrm{T}} \mathbf{x}_i^t$. $\|\cdot\|_1$ is the L_1 norm, which is controlled by μ as a convex approximation to the true vector sparsity. $\|\mathbf{L}\|_F^2$ is the Frobenius norm of matrix $\mathbf{L}$, and λ is the regularization coefficient for matrix $\mathbf{L}$.

Alternating Optimization

If the loss function $\mathcal{L}$ is convex, the objective function in Equation (2.8) is convex in $\mathbf{L}$ for a fixed $\mathbf{S}$, and convex in $\mathbf{S}$ for a fixed $\mathbf{L}$, but they are not jointly convex. Thus, the alternating optimization strategy is adopted to achieve a local minimum. For a fixed $\mathbf{L}$, the optimization function for $\mathbf{s}^t$ becomes:

$$\mathbf{s}^t = \operatorname*{argmin}_{\mathbf{s}^t} \sum_{i=1}^{n_t} \mathcal{L}\left(f(\mathbf{x}_i^t; \mathbf{Ls}^t), y_i^t\right) + \mu \|\mathbf{s}^t\|_1 \quad . \tag{2.9}$$

For a fixed $\mathbf{S}$, the optimization function for $\mathbf{L}$ is:

$$\operatorname*{argmin}_{\mathbf{L}} \sum_{t=1}^{N}\sum_{i=1}^{n_t} \mathcal{L}\left(f(\mathbf{x}_i^t; \mathbf{Ls}^t), y_i^t\right) + \lambda \|\mathbf{L}\|_F^2 \quad . \tag{2.10}$$

For optimization of Equation (2.9), the two-metric projection method in Schmidt et al. [2007] and Gafni and Bertsekas [1984] was used in Kumar et al. [2012]. Equation (2.10) has a closed form solution for squared loss function which is commonly used in regression problems. For classification problems, logistic loss and the Newton-Raphson method were used in optimization in Kumar et al. [2012].

To initialize $\mathbf{L}$, each individual task's parameters are learned independently using their own data, which are stacked as columns in a weight matrix. The top-k left singular vectors of this weight matrix are used as the initial **L**. The reason for this is that the singular vectors are the directions that capture the maximum variances of the task parameters.

2.2.3   DEEP LEARNING IN MULTI-TASK LEARNING

In recent years, DNN has also been applied to MTL. For example, Liu et al. [2015b] proposed a multi-task DNN for learning representations across multiple tasks. They considered two types of tasks: query classification and Web search ranking.

- For query classification, the model classifies whether a query belongs to a particular domain or not. In this work, the authors considered four domains ("Restaurant," "Hotel," "Flight," and "Nightlife"). A query can belong to multiple domains. These four domains are framed as four query classification tasks. The training data for query classification consists of pairs of query and label ($y_t = \{0, 1\}$ where t denotes a particular task or domain).

- For Web search ranking, given a query, the model ranks the documents by their relevance to the query. It is assumed that in the training data, there is at least one relevant document for each query.

In their proposed multi-task DNN model, the lower neural network layers are shared across multiple tasks while the top layers are task-dependent.

The Input Layer (l_1) is the word hash layer in which each word is hashed as a bag of n-grams of letters. For example, the word *deep* is hashed as a bag of letter-trigrams {#-d-e, d-e-e, e-e-p, e-p-#} where # denotes the boundary. This method can hash the variations of the same word into the space close to each other, e.g., *politics* and *politician*.

Semantic-Representation Layer (l_2) maps l_1 into a 300-dimensional vector by:

$$l_2 = f(\mathbf{W}_1 \cdot l_1) \, , \tag{2.11}$$

where $\mathbf{W}_1$ is the weight matrix and f is defined as:

$$f(x) = \frac{1 - e^{-2x}}{1 + e^{-2x}} \, . \tag{2.12}$$

This layer is shared across multiple tasks.

Task-Specific Layer (l_3) for each task converts the 300-dimensional vector to a 128-dimensional vector that is task dependent for each task, using the following:

$$l_3 = f(\mathbf{W}_2^t \cdot l_2) \, , \tag{2.13}$$

where t denotes a particular task and $\mathbf{W}_2^t$ is another weight matrix. For a query classification task, the probability of a query belonging to a domain is obtained from l_3 using a sigmoid function $g(z) = \frac{1}{1+e^{-z}}$. For Web search ranking, the cosine similarity is used to compare layer l_3 of the query and each document. To learn the neural network, the mini-batch-based stochastic gradient descent (SGD) is used, which is an iterative algorithm. In each iteration, a task t is first randomly picked. Then a labeled instance from t is sampled and this labeled instance is used to update the neural network via SGD.

In Collobert and Weston [2008] and Collobert et al. [2011], the authors proposed a unified neural network architecture for performing multiple natural language processing tasks, including part-of-speech tagging, chucking, name entity recognition, semantic role labeling, language modeling, and semantically related words (or synonyms) discovering. They built a DNN for all tasks jointly using weight-sharing. In the neural network, the first layer is for textual features of each word. The second layer extracts features from a sentence, treating the sentence as a sequence rather than a bag of words. The sequence is the input of the second layer. Long-distance dependencies between words in a sentence is captured by Time-Delay Neural Networks (TDNNs) [Waibel et al., 1989], which can model the effects between words beyond a fixed window.

A classic TDNN layer converts a sequence $\mathbf{x}$ to another sequence $\mathbf{o}$ as follows:

$$\mathbf{o}^i = \sum_{j=1-i}^{n-i} \mathbf{L}_j \cdot \mathbf{x}_{i+j} \; , \tag{2.14}$$

where i denotes the time at which the ith word in the sentence is seen in TDNN (i.e., $\mathbf{x}_i$); n is the number of words in the sentence or the length of the sequence. $\mathbf{L}_j$ are the parameters of the layer. Similar to Liu et al. [2015b], stochastic gradient descent is used to train the model, which repeatedly selects a task and one of its training examples to update the neural network.

Along a similar line, Huang et al. [2013a] applied DNN to multilingual data. They proposed a model called shared-hidden-layer multilingual DNN (SHL-MDNN), in which the hidden layers are shared across multiple languages. Furthermore, Zhang et al. [2014] applied deep MTL to the problem of facial landmark detection by co-modeling the correlated tasks such as head pose estimation and facial attribute inference. There are also other applications of deep MTL models to problems such as name error detection in speech recognition [Cheng et al., 2015], multi-label learning [Huang et al., 2013b], phoneme recognition [Seltzer and Droppo, 2013], and so on.

2.2.4   DIFFERENCE FROM LIFELONG LEARNING

The similarity of (batch) MTL and LL is that they both aim to use some shared information across tasks to help learning. The difference is that multi-task learning is still working in the traditional paradigm. Instead of optimizing a single task, it optimizes several tasks simultaneously. If we regard the several tasks as one bigger task, it reduces to the traditional optimization which is actually the case in most optimization formulations of MTL. It does not accumulate any knowledge over time and it not have the concept of continuous learning, which are the key characteristics of LL. Although one can argue that MTL can jointly optimize all tasks whenever a new task is added, it is quite difficult to optimize all tasks in the world simultaneously in a single process as they are too numerous and diverse. Some local and distributed optimizations are needed. Global optimization is also not efficient in terms of both the time and resources. Thus, it is important to retain knowledge to enable incremental learning of multiple tasks with the help of knowledge learned in the past from previous tasks. That is why we regard *online or incremental MTL* as LL.

2.3   ONLINE LEARNING

Online learning is a learning paradigm where the training data points arrive in a sequential order. When a new data point arrives, the existing model is quickly updated to produce the best model so far. Its goal is thus the same as classic learning, i.e., to optimize the performance on the given learning task. It is normally used when it is computationally infeasible to train over the entire dataset or the practical applications cannot wait until a large amount of training data is collected. This is in contrast with the classic batch learning where all training data is available at the beginning for training.

In online learning, if whenever a new data point arrives re-training using all the available data is performed, it will be too expensive. Furthermore, during re-training, the model being used is already out of date. Thus, online learning methods are typically memory and run-time efficient due to the latency requirement in a real-world scenario.

There are a large number of existing online learning algorithms. For example, Kivinen et al. [2004] proposed some online learning algorithms for kernel-based learning like SVM. By extending the classic stochastic gradient descent, they developed computationally efficient online learning algorithms for classification, regression, and novelty detection. Related online kernel classifiers were also studied in Bordes et al. [2005].

Rather than using the traditional table data, Herbster et al. [2005] studied online learning on graphs. Their objective is to learn a function defined on a graph from a set of labeled vertices. One application of their problem is to predict users' preferences toward products in a social network. Ma et al. [2009] worked on the problem of detecting malicious websites using lexical and host-based features and URLs in an online setting. Mairal et al. [2009, 2010] proposed some online dictionary learning approaches for sparse coding, which model data vectors as sparse linear combinations of some basic elements. Hoffman et al. [2010] also proposed an online variational Bayes algorithm for topic modeling.

Much of the online learning research focuses on one domain/task. Dredze and Crammer [2008] developed a multi-domain online learning method, which is based on parameter combination of multiple classifiers. In their setting, the model receives a new instance/example as well as its domain.

2.3.1   DIFFERENCE FROM LIFELONG LEARNING

Although online learning deals with future data in streaming or in a sequential order, its objective is very different from LL. Online learning still performs the same learning task over time. Its objective is to learn more efficiently with the data arriving incrementally. LL, on the other hand, aims to learn from a sequence of different tasks, retain the knowledge learned so far, and use the knowledge to help future task learning. Online learning does not do any of these.

2.4 REINFORCEMENT LEARNING

Reinforcement Learning [Kaelbling et al., 1996, Sutton and Barto, 1998] is the problem where an agent learns actions through trial and error interactions with a dynamic environment. In each interaction step, the agent receives input that contains the current state of the environment. The agent chooses an action from a set of possible actions. The action changes the state of the environment. Then, the agent gets a value of this state transition, which can be reward or penalty. This process repeats as the agent learns a trajectory of actions to optimize its objective. The goal of reinforcement learning is to learn an optimal *policy* that maps states to actions that maximizes the long run sum of rewards. Details about various types of reinforcement learning tasks can be found in Busoniu et al. [2010], Szepesvári [2010], and Wiering and Van Otterlo [2012].

Transfer learning and MTL have also been applied to reinforcement learning. For example, Banerjee and Stone [2007] demonstrated that feature-based value function transfer learning learns optimal policies faster than without knowledge transfer. Taylor et al. [2008] proposed a method to transfer data instances from the source to the target in a model-based reinforcement learning setting. A rule transfer method was also proposed for reinforcement learning [Taylor and Stone, 2007]. An excellent survey of transfer learning applied to reinforcement learning can be found in Taylor and Stone [2009].

Mehta et al. [2008] worked on multiple tasks sharing the same transition dynamics but different reward functions. Instead of fully observable experiments, Li et al. [2009] proposed a model-free multi-task reinforcement learning model for multiple partially observable stochastic environments. They proposed an off-policy batch algorithm to learn parameters in a regionalized policy representation. Lazaric and Ghavamzadeh [2010] assumed that in the multi-task reinforcement learning, only a small number of samples can be generated for any given policy in each task. They grouped the tasks using similar structures and learn them jointly. They also assumed that tasks share structures via value functions which are sampled from a common prior.

Horde, an architecture for learning knowledge in reinforcement learning, was proposed in Sutton et al. [2011]. Its knowledge is represented by a large number of approximate value functions. The reinforcement learning agent is decomposed into many sub-agents. The value function is approximated by the expected return for a trajectory of states and actions. The trajectory is obtained according to the policy of each sub-agent. The intuition is that each sub-agent is responsible for learning some partial information about interactions with the environment. The sub-agents can also use each other's results to achieve their own goals. The final decision of the agent is made by all sub-agents together. However, Sutton et al. [2011] focused on the same environment, which is related to but also different from lifelong learning. Along the lines of Horde, Modayil et al. [2014] modeled a generalization of the value function in reinforcement learning.

2.4.1   DIFFERENCE FROM LIFELONG LEARNING

A reinforcement learning agent learns by trial and error in its interactions with the environment which gives feedback or rewards to the agent. The learning is limited to one task and one environment. There is no concept of accumulating knowledge to help future learning tasks. Transfer and multi-task reinforcement learning paradigms have similar differences from LL as supervised transfer learning and MTL discussed in Sections 2.1.4 and 2.2.4.

2.5   META LEARNING

Meta-learning [Thrun, 1998, Vilalta and Drissi, 2002] primarily aims to learn a new task with only a small number of training examples using a model that has been trained on many other very similar tasks. It is commonly used to solve one-shot or few-shot learning problems. There are usually two learning components in a meta-learning system: a base learner (or a quick learner) and a meta learner (or a slow learner). The base learner is trained within a task with quick updating. The meta learner performs in a task-agnostic meta space, whose goal is to transfer knowledge across tasks. The model learned from the meta learner enables the base learner to learn effectively with only a very small set of training examples. In many cases, the two learners may use the same learning algorithm. With this two-tiered architecture, meta-learning is often described as "learning to learn". But in a nutshell, meta-learning basically treats learning tasks as learning examples. Vilalta and Drissi [2002] gave an excellent overview of the early work on meta-learning. Below, we briefly discuss some more recent papers in the area, particularly in the context of deep neural networks.

Santoro et al. [2016] proposed to consider architectures with augmented memory capacities, such as Neural Turing Machines, to carry short-term and long-term memory demands. An external memory access module was proposed to quickly bind never-before-seen information after few training samples, to improve meta-learning. Andrychowicz et al. [2016] cast the design of optimization algorithms as a learning problem from a meta-learning perspective. A task is defined as a class of problems illustrated by example problem instances. The algorithm was implemented using LSTMs [Hochreiter and Schmidhuber, 1997].

Finn et al. [2016] proposed a model-agnostic meta-learning method that is applicable to any model trained with gradient descent. The key idea is to train the model's initial parameters to fit many tasks well. It is achieved by maximizing the sensitivity of the loss function of new tasks with respect to the parameters. A high sensitivity implies that a small change to the parameters can lead to significant loss amelioration. With such parameters, the model can be simply fine-tuned to perform well on a new task with a small number of training examples. Munkhdalai and Yu [2017] proposed Meta Networks that can update weights at different time-scales. Meta-level information is updated slowly while task-level weights are updated within the scope of each task.

It contains two types of loss functions: a representation loss to create a generalized embedding and a task loss for the specific task. Some other recent works along these lines include [Duan et al., 2017, Grant et al., 2018, Li et al., 2017c, Mishra et al., 2018, Ravi and Larochelle, 2017, Wang et al., 2016, Zhou et al., 2018].

2.5.1 DIFFERENCE FROM LIFELONG LEARNING

Meta-learning trains a meta-model from a large number of tasks to quickly adapt to a new task with only a few examples. One key assumption made by most meta-learning techniques is that the training tasks and test/new tasks are from the same distribution, which is a major weakness and limits the scope of its application. The reason is that in most real-life situations, we would expect that many new tasks have something fundamentally different from old tasks. In the evaluation of meta-learning algorithms, previous tasks are often artificially created to have the same distribution as the new/test tasks. In general, LL does not make this assumption. A lifelong learner is supposed to choose (explicitly or implicitly) the pieces of previous knowledge that are applicable to the new task. If nothing is applicable, no previous knowledge will be used. But clearly, meta-learning is closely related to LL, at least in the aspect of making use of many tasks to help learn the new task. We expect that with further research, the above assumption will be relaxed or even eliminated altogether. In the next edition of this book, we can cover meta-learning fully.

2.6   SUMMARY

In this chapter, we gave an overview of the main ML paradigms that are closely related to LL and described their differences from LL. In summary, we can regard LL as a generalization of or extension to these paradigms. The key characteristics of LL are the continuous learning process, knowledge accumulation in the KB, and the use of past knowledge to help future learning. More advanced features also include learning while working and discovering new problem in applications and learning them in a self-supervised manner based on environmental feedback and previously learned knowledge without manual labeling. The related ML paradigms do not have one or more of these characteristics. In a nutshell, LL essentially tries to mimic the human learning process in order to overcome the limitations of the current isolated learning paradigm. Although we still do not understand the human learning process, that should not prevent us from making progress in ML that exhibits some characteristics of human learning. From the next chapter, we review various existing LL research directions and representative algorithms and techniques.

CHAPTER 3

Lifelong Supervised Learning

This chapter presents existing techniques for lifelong supervised learning (LSL). We first use an example to show why the sharing of information across tasks is useful and how such sharing makes LSL work. The example is about product review sentiment classification. The task is to build a classifier to classify a product review as expressing a positive or negative opinion. In the classic setting, we first label a large number of positive opinion reviews and negative opinion reviews and then run a classification algorithm such as SVM to build a classifier. In the LSL setting, we assume that we have learned from many previous tasks (which may be from different domains). A task here has a set of reviews of a particular kind of product (a *domain*), e.g., camera, cellphone, or car. Let us use the naive Bayesian (NB) classification technique for classifier building. In NB classification, we mainly need the conditional probability of each word w given each class y (positive or negative), $P(w|y)$. When we have a task from a new domain D, the question is whether we need training data from D at all. It is well known that the classifier built in one domain works poorly in another domain because words and language constructs used for expressing opinions in different domains can be quite different [Liu, 2012]. To make matters worse, the same word may express or indicate positive opinion in one domain but negative opinion in another. The answer to the question is *no* in some cases, but *yes* in some others.

The reason for the *no* answer is that we can simply append all training data from the past domains and build a classifier (probably the simplest LL method). This classifier can do wonders for some new domain tasks. It can classify dramatically better than the classifier trained using a modest number of training examples from the new domain D itself. This is because sentiment classification is mainly determined by the words that express positive or negative opinions, called *sentiment words*. For example, *good*, *great*, and *beautiful* are positive sentiment words, and *bad*, *poor*, and *terrible* are negative sentiment words. These words are shared across domains and tasks, but in a particular domain only a small subset of them is used. After seeing the training data from a large number of domains, it is quite clear what words are likely to indicate positive or negative opinions. This means that the system already knows those positive and negative sentiment words and thus can do classification well without any in-domain training reviews from D. To some extent, this is similar to our human case. We don't need a single training positive or negative review to be able to classify reviews into positive and negative classes because we have accumulated so much knowledge in the past about how people praise and criticize things in natural language. Clearly, using one or two past domains for LL is not sufficient because sentiment words used in these domains may be limited and may not even be useful to the new domain. Many non-sentiment words may be regarded as sentiment words incorrectly. Thus, big and diverse data holds a key for LL.

Of course, this simple method does not always work. That is the reason for the *yes* answer above (i.e., requiring some in-domain training data). The reason is that for some domains the sentiment words identified from the past domains can be wrong. For example, the word "toy" usually indicates a negative opinion in a review as people often say that "*this camera is a toy*" and "*this laptop is a toy*." However, when we classify reviews about children's toys, the word "toy" does not indicate any sentiment. We thus need some in-domain training data from D in order to detect such words to overwrite the past knowledge about the words. In fact, this is to solve the problem of *applicability of knowledge* in Section 1.4. With the correction, a lifelong learner can do much better. We will discuss the technique in detail in Section 3.5. In this case, the knowledge base (KB) of LSL stores the empirical counts needed for computing conditional probability $P(w|y)$ in each previous task.

This chapter reviews those representative techniques of LSL. Most of the techniques can perform well with a small number of training examples.

3.1 DEFINITION AND OVERVIEW

We first present the definition of *lifelong supervised learning* (LSL) based on the general definition of lifelong learning (LL) in Chapter 1. We then give a brief overview of the existing work.

Definition 3.1 *Lifelong supervised learning* is a continuous learning process where the learner has performed a sequence of N supervised learning tasks, $\mathcal{T}_1, \mathcal{T}_2, \ldots, \mathcal{T}_N$, and retained the learned knowledge in a knowledge base (KB). When a new task $\mathcal{T}_{N+1}$ arrives, the learner leverages the past knowledge in the KB to help learn a new model f_{N+1} from $\mathcal{T}_{N+1}$'s training data D_{N+1} After learning $\mathcal{T}_{N+1}$, the KB is also updated with the learned knowledge from $\mathcal{T}_{N+1}$.

LSL started with the paper by Thrun [1996b], which proposed several earlier LL methods in the context of memory-based learning and neutral networks. We will review them in Sections 3.2 and 3.3. The neural network approach was then improved in Silver and Mercer [1996, 2002], and Silver et al. [2015]. In these papers, each new task focuses on learning one new concept or class. The goal of LL is to leverage the past data to help build a binary classifier to identify instances of this new class. Ruvolo and Eaton [2013b] proposed the ELLA algorithm to improve the multi-task learning (MTL) method GO-MTL [Kumar et al., 2012] to make it an LL method. Chen et al. [2015] further proposed a technique in the context of NB classification. Clingerman and Eaton [2017] proposed GP-ELLA to support Gaussian processes in ELLA. Xu et al. [2018] presented an LL method for word embedding based on meta-learning. A theoretical study was also conducted by Pentina and Lampert [2014] in the PAC-learning framework. It provided a PAC-Bayesian generalization bound that quantifies the relation between the expected loss on a new task to the average loss on existing tasks for LL. In particular, they modeled the prior knowledge as a random variable and obtained its optimal value by minimizing the expected loss on a new task. Such loss can be transferred from the average loss on existing tasks via the bound. They showed two realizations of the bound on the transfer of parameters [Evgeniou and Pontil, 2004] and the transfer of low-dimensional representations [Ruvolo and Eaton, 2013b]. In the following sections, we present the main existing techniques of LSL.

3.2   LIFELONG MEMORY-BASED LEARNING

In Thrun [1996b], an LSL technique was proposed for two memory-based learning methods: k-nearest neighbors and Shepard's method. We discuss them below.

3.2.1   TWO MEMORY-BASED LEARNING METHODS

K-**Nearest Neighbors** (KNN) [Altman, 1992] is a widely used ML algorithm. Given a testing instance x, the algorithm finds K examples in the training data $\langle x_i, y_i \rangle \in \mathcal{D}$ whose feature vectors x_i are nearest to x according to some distance metric such as the Euclidean distance. The predicted output is the mean value $\frac{1}{K} \sum y_i$ of these nearest neighbors.

Shepard's method [Shepard, 1968] is another commonly used memory-based learning method. Instead of only using K examples as in KNN, this method uses all the training examples in $\mathcal{D}$ and weights each example according to the inverse distance to the test instance x, as shown below:

$$ s(x) = \left(\sum_{\langle x_i, y_i \rangle \in \mathcal{D}} \frac{y_i}{\|x - x_i\| + \epsilon} \right) \times \left(\sum_{\langle x_i, y_i \rangle \in \mathcal{D}} \frac{1}{\|x - x_i\| + \epsilon} \right)^{-1}, \tag{3.1} $$

where $\epsilon > 0$ is a small constant to avoid the denominator being zero. Neither KNN nor Shepard's method can use the previous task data with different distributions or distinct class labels to help its classification.

3.2.2   LEARNING A NEW REPRESENTATION FOR LIFELONG LEARNING

Thrun [1996b] proposed to learn a new representation to bridge the gap among tasks for the above two memory-based methods to achieve LL, which was shown to improve the predictive performance especially when the number of labeled examples is small.

The interest of the paper is concept learning. Its goal is to learn a function $f : I \rightarrow \{0, 1\}$ where $f(x) = 1$ means that $x \in I$ belongs to a target concept (e.g., *cat* or *dog*); otherwise x does not belong to the concept. For example, $f_{dog}(x) = 1$ means that x is an instance of the concept dog. Let the data from the previous N tasks be $\mathcal{D}^p = \{\mathcal{D}_1, \mathcal{D}_2, \ldots, \mathcal{D}_N\}$. Each past task data $\mathcal{D}_i \in \mathcal{D}^p$ is associated with an unknown classification function f_i. $\mathcal{D}^p$ is called the *support set* in Thrun [1996b]. The goal is to learn the function f_{N+1} for the current new task data $\mathcal{D}_{N+1}$ with the help of the support set.

To bridge the difference among different tasks and to be able to exploit the shared information in the past data (the support set), the paper proposed to learn a new representation of the data, i.e., to learn a *space transformation function* $g : I \rightarrow I'$ to map the original input feature vectors in I to a new space I'. The new space I' then serves as the input space for KNN or the Shepard's method. The intuition is that positive examples of a concept (with $y = 1$) should have similar new representations while a positive example and a negative example of a concept ($y = 1$ and $y = 0$) should have very different representations. This idea can be formulated as an energy function E for g:

$$E = \sum_{\mathcal{D}_i \in \mathcal{D}^p} \sum_{\langle x, y=1 \rangle \in \mathcal{D}_i} \left(\sum_{\langle x', y'=1 \rangle \in \mathcal{D}_i} \| g(x) - g(x') \| - \sum_{\langle x', y'=0 \rangle \in \mathcal{D}_i} \| g(x) - g(x') \| \right) . \quad (3.2)$$

The optimal function g^* is achieved by minimizing the energy function E, which forces the distance between pairs of positive examples of the concept ($\langle x, y = 1 \rangle$ and $\langle x', y' = 1 \rangle$) to be small, and the distance between a positive example and a negative example of a concept ($\langle x, y = 1 \rangle$ and $\langle x', y' = 0 \rangle$) to be large. In the implementation of Thrun [1996b], g was realized with a neural network and trained with the support set using Back-Propagation.

Given the mapping function g^*, rather than performing memory-based learning in the original space $\langle x_i, y_i \rangle \in \mathcal{D}_{N+1}$, x_i is first transformed to the new space using g^* to $\langle g^*(x_i), y_i \rangle$ before applying KNN or the Shepard's method.

Since this approach does not retain any knowledge learned in the past but only accumulates the past data, it is thus inefficient if the number of previous tasks is large because the whole optimization needs to be re-done using all the past data (the support set) whenever a new task arrives. In Thrun [1996b], an alternative method to the above energy-function-based approach was also proposed, which learns a distance function based on the support set. This distance function is then used in lifelong memory-based learning. This approach has similar weaknesses. These techniques also do not deal with the correctness or applicability of the shared knowledge g^* (Section 1.4).

3.3 LIFELONG NEURAL NETWORKS

Here we introduce two early neural network approaches to LSL.

3.3.1 MTL NET

Although MTL net (Multi-task learning with neural network) [Caruana, 1997] is described as an LL method in Thrun [1996b], it is actually a batch MTL method. Based on our definition of LL, they are different learning paradigms. However, for historical reasons, we still give it a brief discussion here.

In MTL net, instead of building a neural network for each individual task, it constructs a universal neural network for all the tasks (see Figure 3.1). This universal neural network uses the same input layer for input from all tasks and uses one output unit for each task (or class in this case). There is also a shared hidden layer in MTL net that is trained in parallel using Back-Propagation [Rumelhart et al., 1985] on all the tasks to minimize the error on all the tasks. This shared layer allows features developed for one task to be used by other tasks. So some developed features can represent the common characteristics of the tasks. For a specific task, it will activate some hidden units that are related to it while making the weights of the other irrelevant hidden units small. Essentially, like a normal batch MTL method, the system jointly optimizes the classification of all the past/previous and the current/new tasks. It is thus not regarded as an LL method based on the definition in this book (see Section 2.2.4). Several extensions of MTL net were made in Silver and Mercer [2002], Silver and Poirier [2004, 2007], from generating and using virtual training examples to deal with the need for the training data of all previous tasks to adding contexts.

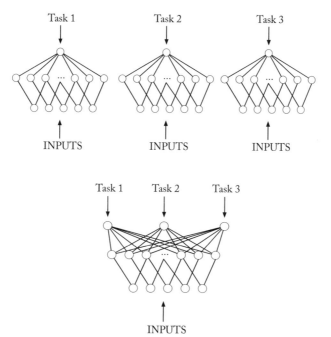

Figure 3.1: The top neural networks are trained independently for each task, and the bottom neural network is MTL net [Caruana, 1997].

3.3.2   LIFELONG EBNN

This LL approach is in the context of EBNN (Explanation-Based Neural Network) [Thrun, 1996a], which again leverages the previous task data (or the support set) to improve learning. As in Section 3.2.2, concept learning is the goal of this work, which learns a function $f : I \rightarrow \{0, 1\}$

to predict if an object represented by a feature vector $x \in I$ belongs to a concept ($y = 1$) or not ($y = 0$).

In this approach, the system first learns a *general distance function*, $d : I \times I \rightarrow [0, 1]$, considering all the past data (or the support set) and uses this distance function to share or transfer the knowledge of the past task data to the new task $\mathcal{T}_{N+1}$. Given two input vectors, say x and x', function d computes the probability of x and x' being members of the same concept (or class), regardless what the concept is. In Thrun [1996b], d is learned using a neural network trained with Back-Propagation. The training data for learning the distance function is generated as follows: For each past task data $\mathcal{D}_i \in \mathcal{D}^P$, each pair of examples of the concept generates a training example. For a pair, $\langle x, y = 1 \rangle \in \mathcal{D}_i$ and $\langle x', y' = 1 \rangle \in \mathcal{D}_i$, a positive training example is generated, $\langle (x, x'), 1 \rangle$. For a pair $\langle x, y = 1 \rangle \in \mathcal{D}_i$ and $\langle x', y' = 0 \rangle \in \mathcal{D}_i$ or $\langle x, y = 0 \rangle \in \mathcal{D}_i$ and $\langle x', y' = 1 \rangle \in \mathcal{D}_i$, a negative training example is generated, $\langle (x, x'), 0 \rangle$.

With the learned distance function in hand, EBNN works as follows: Unlike a traditional neural network, EBNN estimates the *slope* (tangent) of the target function at each data point x and adds it into the vector representation of the data point. In the new task $\mathcal{T}_{N+1}$, a training example is of the form, $\langle x, f_{N+1}(x), \nabla_x f_{N+1}(x) \rangle$, where $f_{N+1}(x)$ is just the original class label of $x \in \mathcal{D}_{N+1}$ (the new task data). The system is trained using Tangent-Prop algorithm [Simard et al., 1992]. $\nabla_x f_{N+1}(x)$ is estimated using the gradient of the distance d obtained from the neural network, i.e., $\nabla_x f_{N+1}(x) \approx \frac{\partial d_{x'}(x)}{\partial x}$, where $\langle x', y' = 1 \rangle \in \mathcal{D}_{N+1}$ and $d_{x'}(x) = d(x, x')$. The rationale is that the distance between x and a positive training example x' is an estimate of the probability of x being a positive example, which approximates $f_{N+1}(x)$. As a result, the built EBNN fits both the current task data $\mathcal{D}_{N+1}$ and the support set through $\nabla_x f_{N+1}(x)$ and d.

Similar to lifelong KNN in Section 3.2, in this case, the part of the system that learns the distance function (the *shared knowledge*) and performs EBNN is the *knowledge-based learner* in Section 1.4. Again, the *knowledge base* stores only the past data. Similarly, this technique also does not deal with correctness or applicability of the shared knowledge d (see Section 1.4).

Like lifelong KNN, since lifelong EBNN does not retain any knowledge learned in the past but only accumulates the past data, it is also inefficient if the number of previous tasks is large because the training of the distance function d needs to be re-done using all the past data (the support set) whenever a new task arrives. Additionally, since every pair of data points in each past task dataset forms a training example for learning the distance function d, the training data for learning d can be huge.

3.4   ELLA: AN EFFICIENT LIFELONG LEARNING ALGORITHM

This section focuses on the lifelong supervised learning (LSL) system ELLA (Efficient Lifelong Learning Algorithm) proposed by Ruvolo and Eaton [2013a,b]. It maintains a sparsely shared basis (the past knowledge) for all task models, transfers knowledge from the basis to the new task, and refines the basis over time to maximize the performances across all tasks. Unlike cumulative learning, each task in ELLA is independent of other tasks. ELLA also follows the tradition of MTL aimed at optimizing the performances of all tasks. Many other LL methods mainly optimize the performance of the new task, although they can help optimize any previous task if needed. In the presentation below, we try to use about the same notation as in the original paper for easy reference.

3.4.1 PROBLEM SETTING

As in a normal LL problem, ELLA receives a sequence of supervised learning tasks, $1, 2, \ldots, N$, in a lifelong manner. Each task t has its training data $\mathcal{D}^t = \{(\mathbf{x}_i^t, y_i^t) : i = 1, \ldots, n_t\}$, where n_t is the number of training instances in $\mathcal{D}^t$, and is defined by a hidden (or latent) true mapping $\hat{f}^t(\mathbf{x})$ from an instance space $\mathcal{X}^t \subseteq \mathbb{R}^d$ to a set of labels $\mathcal{Y}^t$ (or $\mathcal{Y}^t = \mathbb{R}$ for regression). Let d be the feature dimension.

ELLA extends the batch MTL model GO-MTL [Kumar et al., 2012] (also in Section 2.2.2) to make it more efficient and become an incremental or online MTL system, which is regarded as an LL system. Like Go-MTL, ELLA takes a parametric approach to model building in which the model or the prediction function $f^t(\mathbf{x}) = f^t(\mathbf{x}; \boldsymbol{\theta}^t)$ for each task t is governed by a task-specific parameter vector $\boldsymbol{\theta}^t \in \mathbb{R}^d$. The goal of ELLA is to construct task models $f^1, \ldots, f^N$ so that:

1. for each task, $f^t \approx \hat{f}^t$;

2. a new model f^t can be added quickly when the training data for a new task t arrives; and

3. each past model f^t can be updated efficiently after the addition of the new task.

ELLA assumes that the total number of tasks, the distribution of tasks and their order are all unknown [Ruvolo and Eaton, 2013b]. It also assumes that there can be a large number of tasks, while each task can have a large number of data points. Thus, an LL algorithm that is both effective and efficient is needed.

3.4.2 OBJECTIVE FUNCTION

In the same way as the GO-MTL model [Kumar et al., 2012] (see Section 2.2.2), ELLA maintains k sparsely shared basis model components for all task models. Let $\mathbf{L} \subseteq \mathbb{R}^{d \times k}$ be the k basis model components. Each task model's parameter vector $\boldsymbol{\theta}^t$ is assumed to be a linear combination of the weight vector $\mathbf{s}^t \in \mathbb{R}^k$ and the basis model components $\mathbf{L}$. We thus obtain the equation below (same as Equation (2.7)):

$$\underset{d \times N}{\boldsymbol{\Theta}} = \underset{d \times k}{\mathbf{L}} \times \underset{k \times N}{\mathbf{S}}, \tag{3.3}$$

where $\boldsymbol{\Theta} = [\boldsymbol{\theta}^1, \boldsymbol{\theta}^2, \ldots, \boldsymbol{\theta}^N]$ and $\mathbf{S} = [\mathbf{s}^1, \mathbf{s}^2, \ldots, \mathbf{s}^N]$. For each task t, $\boldsymbol{\theta}^t = \mathbf{L}\mathbf{s}^t$. The initial objective function of ELLA is the same as Equation (2.8) in GO-MTL except that it optimizes the average (rather than the sum) loss on the training data across all tasks, which is essential for the convergence guarantees:

$$\frac{1}{N} \sum_{t=1}^{N} \min_{s^t} \left\{ \frac{1}{n_t} \sum_{i=1}^{n_t} \mathcal{L} \left(f(\mathbf{x}_i^t; \mathbf{L}\mathbf{s}^t), y_i^t \right) + \mu \|\mathbf{s}^t\|_1 \right\} + \lambda \|\mathbf{L}\|_F^2 \quad , \tag{3.4}$$

where $f(\mathbf{x}_i^t; \mathbf{L}\mathbf{s}^t) = \boldsymbol{\theta}^t \mathbf{x}_i^t = (\mathbf{L}\mathbf{s}^t)^\top \mathbf{x}_i^t$. Since the objective function is not jointly convex in $\mathbf{L}$ and the $\mathbf{s}^t$'s, to optimize it, one can use a common approach to computing a local optimum, i.e., alternately optimizing $\mathbf{s}^t$ while holding $\mathbf{L}$ fixed, and optimizing $\mathbf{L}$ while holding $\mathbf{s}^t$ fixed. However, as pointed out in Ruvolo and Eaton [2013b], there are two major inefficiency issues in the above objective function (which also exist in GO-MTL).

1. There is an explicit dependence on *all* of the previous training data (through the inner summation). That is, to compute the objective function, one needs to iterate all training instances in all tasks in order to compute their loss function values. If the number of tasks is large or the number of training instances in each task is large, this iteration can be very inefficient.

2. When evaluating a single candidate $\mathbf{L}$ in Equation (3.4), an optimization problem must be solved to recompute the value of each $\mathbf{s}^t$. This means each $\mathbf{s}^t$ will have to be updated when $\mathbf{L}$ is updated. This becomes increasingly expensive when there are more and more tasks.

Ruvolo and Eaton [2013b] proposed some approximation techniques to deal with the above two inefficiency issues, which we detail in the next subsection. The basic idea is to approximate the fit of a new task model using the single-task solution as a point estimate in the basis of model components learned over the past tasks, and then updates the basis to incorporate the new knowledge from the new task.

3.4.3 DEALING WITH THE FIRST INEFFICIENCY

To address the first issue, Ruvolo and Eaton [2013b] used the second-order Taylor expansion for approximation. Before giving the technical details, let us briefly review some mathematical foundations.

Taylor Expansion
In the single-variable case, i.e., when $g(x)$ is a one variable function, the second-order Taylor expansion near a constant value a is:

$$g(x) \approx g(a) + g'(a)(x - a) + \frac{1}{2} g''(a)(x - a)^2 \quad , \tag{3.5}$$

where $g'()$ and $g''()$ are the derivative and the second-order derivative of function g.

In the multiple-variable case, i.e., when $g(\mathbf{x})$ is a multivariate function (assuming $\mathbf{x}$ has n values), the second-order Taylor expansion near vector $\mathbf{a}$ of a constant size n is:

$$g(\mathbf{x}) \approx g(\mathbf{a}) + \nabla g(\mathbf{a})(\mathbf{x} - \mathbf{a}) + \frac{1}{2} \|(\mathbf{x} - \mathbf{a})\|_{\boldsymbol{H}(\mathbf{a})}^2 \ , \tag{3.6}$$

where $\|\mathbf{v}\|_{\mathbf{A}}^2 = \mathbf{v}^{\mathrm{T}} \mathbf{A} \mathbf{v}$ and $\boldsymbol{H}(\mathbf{a})$ is called *Hessian Matrix* of function g.

Optimality Conditions in Unconstrained Optimization

Consider the problem of minimizing function $f : \mathbb{R}^n \to \mathbb{R}$, where f is twice continuously differentiable on $\mathbb{R}^n$:

$$\min_{\mathbf{x} \in \mathbb{R}^n} f(\mathbf{x}) \ . \tag{3.7}$$

Theorem 3.2 First-Order Necessary Conditions for Optimality. *Let function $f : \mathbb{R}^n \to \mathbb{R}$ be differentiable at a point $\hat{\mathbf{x}} \in \mathbb{R}^n$. If $\hat{\mathbf{x}}$ is a local minimizer, then $\nabla f(\hat{\mathbf{x}}) = 0$.*

Proof. From the definition of the first-order Taylor expansion, we have:

$$f(\mathbf{x}) = f(\hat{\mathbf{x}}) + \nabla f(\hat{\mathbf{x}})^{\mathrm{T}} (\mathbf{x} - \hat{\mathbf{x}}) + o(\|\mathbf{x} - \hat{\mathbf{x}}\|) \ ; \tag{3.8}$$

that is,

$$f(\mathbf{x}) - f(\hat{\mathbf{x}}) = \nabla f(\hat{\mathbf{x}})^{\mathrm{T}} (\mathbf{x} - \hat{\mathbf{x}}) + o(\|\mathbf{x} - \hat{\mathbf{x}}\|) \ , \tag{3.9}$$

where $\lim_{\mathbf{x} \to \hat{\mathbf{x}}} \frac{o(\|\mathbf{x} - \hat{\mathbf{x}}\|)}{\|\mathbf{x} - \hat{\mathbf{x}}\|} = 0$. Let $\mathbf{x} := \hat{\mathbf{x}} - \alpha \nabla f(\hat{\mathbf{x}})$, where α is a positive constant. Plugging it into Equation (3.9), then:

$$0 \le \frac{f(\hat{\mathbf{x}} - \alpha \nabla f(\hat{\mathbf{x}})) - f(\hat{\mathbf{x}})}{\alpha} = -\|\nabla f(\hat{\mathbf{x}})\|^2 + \frac{o(\alpha \|\nabla f(\hat{\mathbf{x}})\|)}{\alpha} \ . \tag{3.10}$$

Taking the limit as $\alpha \downarrow 0$, we obtain:

$$0 \le -\|\nabla f(\hat{\mathbf{x}})\|^2 \le 0 \ . \tag{3.11}$$

Hence, $\nabla f(\hat{\mathbf{x}}) = 0$. $\qquad\qquad\square$

Removing Dependency

We now come back to ELLA. To remove the explicit dependence on *all* task training data, the second-order Taylor expansion is used to approximate the objective function in Equation (3.4). Let's first define a function $g(\boldsymbol{\theta}^t)$ as below:

$$g(\boldsymbol{\theta}^t) = \frac{1}{n_t} \sum_{i=1}^{n_t} \mathcal{L}\left(f(\boldsymbol{x}_i^t; \boldsymbol{\theta}^t), y_i^t\right) \ , \tag{3.12}$$

where $\boldsymbol{\theta}^t = \mathbf{L}\mathbf{s}^t$. Then the objective function in Equation (3.4) becomes:

$$\frac{1}{N} \sum_{t=1}^{N} \min_{\mathbf{s}^t} \left\{ g(\boldsymbol{\theta}^t) + \mu \|\mathbf{s}^t\|_1 \right\} + \lambda \|\mathbf{L}\|_F^2 \quad . \tag{3.13}$$

Let's assume that the minimum solution of the function g is $\hat{\boldsymbol{\theta}}^t$, i.e., $\hat{\boldsymbol{\theta}}^t = \text{argmin}_{\boldsymbol{\theta}^t} \frac{1}{n_t} \sum_{i=1}^{n_t} \mathcal{L}\left(f(\mathbf{x}_i^t; \boldsymbol{\theta}^t), y_i^t\right)$ (which is an optimal predictor learned on only the training data for task t). Then, the second-order Taylor expansion near $\hat{\boldsymbol{\theta}}^t$ is as follows:

$$g(\boldsymbol{\theta}^t) \approx g(\hat{\boldsymbol{\theta}}^t) + \nabla g(\hat{\boldsymbol{\theta}}^t)(\boldsymbol{\theta}^t - \hat{\boldsymbol{\theta}}^t) + \frac{1}{2}\|\boldsymbol{\theta}^t - \hat{\boldsymbol{\theta}}^t\|_{\boldsymbol{H}^t}^2 \quad , \tag{3.14}$$

where $\boldsymbol{H}^t = \boldsymbol{H}(\hat{\boldsymbol{\theta}}^t)$ is the Hessian Matrix of function g.

Considering that function g is used in the outer minimization in Equation (3.13), the first constant term in Equation (3.14) can be suppressed. According to the first-order necessary conditions (Theorem 3.2), $\nabla g(\hat{\boldsymbol{\theta}}^t) = 0$ since $\hat{\boldsymbol{\theta}}^t$ is the local minimum solution of function g, and thus the second term in Equation (3.14) can also be removed. Hence, the new objective function after plugging in Equation (3.13) is:

$$\frac{1}{N} \sum_{t=1}^{N} \min_{\mathbf{s}^t} \left\{ \|\boldsymbol{\theta}^t - \hat{\boldsymbol{\theta}}^t\|_{\boldsymbol{H}^t}^2 + \mu \|\mathbf{s}^t\|_1 \right\} + \lambda \|\mathbf{L}\|_F^2 \quad . \tag{3.15}$$

As $\boldsymbol{\theta}^t = \mathbf{L}\mathbf{s}^t$, Equation (3.15) can be rewritten as:

$$\frac{1}{N} \sum_{t=1}^{N} \min_{\mathbf{s}^t} \left\{ \|\hat{\boldsymbol{\theta}}^t - \mathbf{L}\mathbf{s}^t\|_{\boldsymbol{H}^t}^2 + \mu \|\mathbf{s}^t\|_1 \right\} + \lambda \|\mathbf{L}\|_F^2 \quad . \tag{3.16}$$

$$\boldsymbol{H}^t = \frac{1}{2} \nabla_{\boldsymbol{\theta}^t, \boldsymbol{\theta}^t}^2 \frac{1}{n_t} \sum_{i=1}^{n_t} \mathcal{L}\left(f(\mathbf{x}_i^t; \boldsymbol{\theta}^t), y_i^t\right) \bigg|_{\boldsymbol{\theta}^t = \hat{\boldsymbol{\theta}}^t} \quad , \quad \text{and}$$

$$\hat{\boldsymbol{\theta}}^t = \text{argmin}_{\boldsymbol{\theta}^t} \frac{1}{n_t} \sum_{i=1}^{n_t} \mathcal{L}\left(f(\mathbf{x}_i^t; \boldsymbol{\theta}^t), y_i^t\right) \quad .$$

Note that $\hat{\boldsymbol{\theta}}^t$ and $\boldsymbol{H}^t$ will remain the same if the training data for task t does not change. Thus, the new objective function in Equation (3.16) removes the dependence of the optimization on the training data of all previous tasks.

3.4.4   DEALING WITH THE SECOND INEFFICIENCY

The second efficiency issue is that when computing a single candidate $\mathbf{L}$, an optimization problem must be solved to recompute the value of each $\mathbf{s}^t$. To solve this problem, Ruvolo and Eaton

[2013b] adopted this strategy: when the training data for task t is last encountered, only s^t is updated while $s^{t'}$ for other tasks t' remain the same. That is, s^t is computed when the training data for task t is last encountered, and it is not updated later when training on other tasks. Although this seems to prevent the influence of earlier tasks from later tasks, they will benefit from the subsequent adjustment of the basis latent model components $\mathbf{L}$. Using the previously computed values of s^t, the following optimization process is performed:

$$s^t \leftarrow \operatorname*{argmin}_{s^t} \|\hat{\boldsymbol{\theta}}^t - \mathbf{L}_m s^t\|^2_{\boldsymbol{H}^t} + \mu \|s^t\|_1 \,, \text{ with fixed } \mathbf{L}_m, \text{ and}$$

$$\mathbf{L}_{m+1} \leftarrow \operatorname*{argmin}_{\mathbf{L}} \frac{1}{N} \sum_{t=1}^{N} \left(\|\hat{\boldsymbol{\theta}}^t - \mathbf{L} s^t\|^2_{\boldsymbol{H}^t} + \mu \|s^t\|_1 \right) + \lambda \|\mathbf{L}\|^2_F, \text{ with fixed } s^t \,,$$

where notation $\mathbf{L}_m$ refers to the value of the latent components at the mth iteration and t is assumed to be the particular task for which the training data just arrives. Note that if t is an existing task, the new training data is merged into the old training data of t.

For the specific steps in performing the updates in the preceding equations, please refer to the original paper. They depend on the type of model and the loss function used. The paper presented two cases: linear regression and logistic regression.

3.4.5   ACTIVE TASK SELECTION

LL in the above problem setting (Section 3.4.1) is a passive process, i.e., the system has no control over the order in which the learning tasks are presented. Ruvolo and Eaton [2013a] considered ELLA in an active task selection setting. Assuming that there is a pool of candidate tasks, rather than choosing a task randomly as in ELLA, Ruvolo and Eaton [2013a] chose tasks in a certain order with the purpose of maximizing future learning performance using as few tasks as possible. The problem is practical since each learning task may need a significant amount of time of manual labeling or each learning task may take a long time for the system to run. In such cases, learning in a task-efficient manner by choosing some tasks in certain order is more scalable to real-life LL problems.

Active Task Selection Setting

The active task selection setting in LL is defined as follows: instead of modeling training data of task t as in regular LL, the system has a pool of candidate unlearned tasks $\mathcal{T}_{pool}$ to choose from. For each candidate task $t \in \mathcal{T}_{pool}$, only a subset of training instances is labeled, which are denoted by $\mathcal{D}^t_c = (\mathbf{X}^t_c, \mathbf{Y}^t_c)$. Based on these small subsets, one of the tasks, $t_{next} \in \mathcal{T}_{pool}$, is chosen to learn next. After that, all the training data of t_{next} will be revealed, which is denoted by $\mathcal{D}^{(t_{next})} = (\mathbf{X}^{(t_{next})}, \mathbf{Y}^{(t_{next})})$. Note that for each task t, $\mathcal{D}^t_c \subseteq \mathcal{D}^t$. The size of the candidate pool can be a fixed value or increase/decrease dynamically during learning.

Diversity Method

Here we introduce the *diversity* method for active task selection proposed in Ruvolo and Eaton [2013a] which was shown to perform the best compared to the other methods used in the paper. In the context of ELLA, in order to maximize performance on future tasks, the model should have a flexible and robust set of latent components, i.e., $\mathbf{L}$. In other words, $\mathbf{L}$ should be adaptable to a wide variety of tasks. If $\mathbf{L}$ does not fit well for a new task t, it means that the information in t has not been represented well in the current $\mathbf{L}$. Thus, in order to solve the widest range of tasks, the next task should be the one that the current basis $\mathbf{L}$ performs the worst, i.e., the loss on the subset of the labeled data is maximal. This heuristic is described as follows:

$$t_{next} = \underset{t \in \mathcal{T}_{pool}}{\operatorname{argmax}} \ \min_{\mathbf{s}^t} \|\hat{\boldsymbol{\theta}}^t - \mathbf{L}\mathbf{s}^t\|_{\boldsymbol{H}^t}^2 + \mu\|\mathbf{s}^t\|_1 \quad , \tag{3.17}$$

where $\hat{\boldsymbol{\theta}}^t$ and $\boldsymbol{H}^t$ are obtained from the subset of the labeled data $\mathcal{D}_c^t$. Since Equation (3.17) tends to select tasks that are encoded poorly with the current basis $\mathbf{L}$, the selected tasks are likely to be very different from existing tasks, and it thus encourages diverse tasks.

Rather than simply choosing the task with the maximal loss value, another way (called *Diversity++*) is to estimate the probability of selecting task t as the square value of the minimal loss value for t, as below:

$$p(t_{next} = t) \propto \left(\min_{\mathbf{s}^t} \|\hat{\boldsymbol{\theta}}^t - \mathbf{L}\mathbf{s}^t\|_{\boldsymbol{H}^t}^2 + \mu\|\mathbf{s}^t\|_1 \right)^2 . \tag{3.18}$$

Then each time, a task is sampled based on the probability $p(t_{next})$. This is thus a stochastic variant of the diversity method above.

3.5   LIFELONG NAIVE BAYESIAN CLASSIFICATION

This section introduces the *lifelong NB classification* technique given in Chen et al. [2015]. It is applied to a sentiment analysis task, classifying whether a product review expresses a positive or negative opinion. The system is called LSC (lifelong sentiment classification). Below we first briefly introduce the NB classification formulation and then introduce its lifelong extension for sentiment classification. Again for easy reference, we follow the notation in the original paper.

3.5.1   NAÏVE BAYESIAN TEXT CLASSIFICATION

NB for text classification is a generative model consisting of a mixture of multinomial distributions. Each multinomial distribution (called a mixture component) is the generator of a single class of documents. Training an NB model is to find the parameters of each multinomial distribution and the mixture weight. For text classification, the above idea can be translated as follows: Given a set of training documents $\mathcal{D} = \{d_1, d_2, \ldots, d_{|\mathcal{D}|}\}$, a vocabulary of V (the set of distinct words/terms in $\mathcal{D}$) and a set of classes $C = \{c_1, c_2, \ldots, c_{|C|}\}$ associated with $\mathcal{D}$, NB classification trains a classifier by computing the conditional probability of each word $w \in V$ given each class c_j, i.e., $P(w|c_j)$ (the model parameter for class c_j) and the prior probability of each class, $P(c_j)$ (the mixture weight) [McCallum and Nigam, 1998].

$P(w|c_j)$ is estimated based on the empirical word counts as follows:

$$P\left(w|c_j\right) = \frac{\lambda + N_{c_j,w}}{\lambda |V| + \sum_{v=1}^{|V|} N_{c_j,v}} \quad , \tag{3.19}$$

where N_{c_j} is the number of times that word w occurs in the documents of class c_j. λ ($0 \leq \lambda \leq 1$) is used for smoothing. When $\lambda = 1$, it is known as *Laplace smoothing*. The prior probability of each class, $P(c_j)$, is estimated as follows:

$$P(c_j) = \frac{\sum_{i=1}^{|\mathcal{D}|} P(c_j|d_i)}{|\mathcal{D}|} \quad , \tag{3.20}$$

where $P(c_j|d_i) = 1$ if c_j is the label of the training document d_i and 0 otherwise.

For testing, given a test document d, NB computes the posterior probability $P(c_j|d)$ for each class c_j and picks the class with the highest $P(c_j|d)$ as the classification result:

$$P(c_j|d) = \frac{P(c_j)P(d|c_j)}{P(d)} \tag{3.21}$$

$$= \frac{P(c_j)\prod_{w \in d} P(w|c_j)^{n_{w,d}}}{\sum_{r=1}^{|C|} P(c_r)\prod_{w \in d} P(w|c_r)^{n_{w,d}}} \quad , \tag{3.22}$$

where $n_{w,d}$ is the number of times that word w appears in d.

NB is a natural fit for LL because past knowledge can serve as priors for the probabilities of the new task very easily. LSC exploits this idea. Let us answer two specific questions in the context of sentiment classification. The first question is why the past learning can contribute to the new/current task classification given that the current task already has labeled training data. The answer is that the training data may not be fully representative of the test data due to *sample selection bias* [Heckman, 1979, Shimodaira, 2000, Zadrozny, 2004] and/or small training data size, which is the case in Chen et al. [2015]. For example, in a sentiment classification application, the test data may contain some sentiment words that are absent in the current training data, but they have appeared in review data in some previous tasks. So the past knowledge can provide the prior sentiment polarity information for the current new task. Note that for sentiment classification, sentiment words such as *good*, *nice*, *terrible*, and *poor* are instrumental. Note also that each task in Chen et al. [2015] is actually from a different domain (or products). We thus use *task* and *domain* interchangeably from now on.

The second question is why the past knowledge can help even if the past domains are very diverse and not very similar to the current domain. The main reason is that in sentiment classification, sentiment words and expressions are largely domain independent. That is, their polarities (positive or negative) are often shared across domains. Hence having worked a large number of previous/past domains, the system has learned a lot of positive and negative sentiment words. It is important to note that only one or two past domains are not sufficient because of the low coverage of sentiment words in the limited domains.

3.5.2   BASIC IDEAS OF LSC

This subsection introduces the basic ideas of the LSC technique. We start by discussing what is stored in the knowledge base of LSC.

Knowledge Base

For each word $w \in V^p$ (where V^p is the vocabulary of all previous tasks), the knowledge base *KB* stores two types of information: *document-level knowledge* and *domain-level knowledge*.

1. *Document-level knowledge* $N_{+,w}^{KB}$ (and $N_{-,w}^{KB}$): number of occurrences of w in the documents of the positive (and negative) class in the previous tasks.

2. *Domain-level knowledge* $M_{+,w}^{KB}$ (and $M_{-,w}^{KB}$): number of domains in which $P(w|+) > P(w|-)$ (and $P(w|+) < P(w|-)$). Here, in each previous task, $P(w|+)$ and $P(w|-)$ are calculated using Equation (3.19). Here $+$ and $-$ stands for positive and negative opinion classes respectively.

The domain-level knowledge is complementary to the document-level knowledge as w may be extremely frequent in a domain but rare in other domains which leads to the superfluous effect of that domain on w at the document level.

A Naïve Approach to Using Knowledge

From Section 3.5.1, we can see that the key parameters that affect NB classification results are $P(w|c_j)$ which are computed using the empirical counts $N_{c_j,w}$ and the total number of words in the class of documents. In binary classification, $P(w|c_j)$ are computed using $N_{+,w}$ and $N_{-,w}$. This suggests that we can revise these counts appropriately to improve classification. Given the new task data $\mathcal{D}^t$, we denote the empirical word counts $N_{+,w}^t$ (and $N_{-,w}^t$) as the number of times that word w occurs in the positive (and negative) documents in $\mathcal{D}^t$. Here, we explicitly use superscript t to distinguish it from the previous tasks. The task becomes how to effectively use the knowledge in the *KB* to update word counts to build a superior NB classifier.

Given the knowledge base *KB* from the past learning tasks, one naïve way to build a classifier is to sum up the counts in *KB* (served as priors) with the empirical counts $N_{+,w}^t$ and $N_{-,w}^t$ of $\mathcal{D}^t$, i.e., $X_{+,w} = N_{+,w}^t + N_{+,w}^{KB}$ and $X_{-,w} = N_{-,w}^t + N_{-,w}^{KB}$. Here, we call $X_{+,w}$ and $X_{-,w}$ *virtual counts* as they will be updated using optimization discussed in the next sub-section. In building the classifier, $N_{+,w}$ and $N_{-,w}$ (i.e., $N_{c_j,w}$) in Equation (3.19) are replaced by $X_{+,w}$ and $X_{-,w}$, respectively. This naïve method turns out to be quite good in many cases, but it has two weaknesses.

1. The past domains usually contain much more data than the current domain, which means $N_{+,w}^{KB}$ (and $N_{-,w}^{KB}$) may be much larger than $N_{+,w}^{t}$ (and $N_{-,w}^{t}$). As a result, the merged results may be dominated by the counts in the *KB* from the past domains.

2. It does not consider the domain-dependent word polarity. A word may be positive in the current domain but negative in past domains. For example, past domains may indicate that the word "toy" is negative because there are a lot of past sentences like "this product is a toy." However, in the toy domain, the word expresses no sentiment.

The LSC system solves these two problems using an optimization method.

3.5.3   LSC TECHNIQUE

LSC uses *stochastic gradient descent* (SGD) to minimize the training error by adjusting $X_{+,w}$ and $X_{-,w}$ (virtual counts), which are the numbers of times that a word w appears in the positive and negative classes, respectively.

For correct classification, ideally, we should have the posterior probability $P(+|d_i) = 1$ for each positive class (+) document d_i, and $P(-|d_i) = 1$ for each negative class (−) document d_i. In stochastic gradient descent, we optimize the classification of each $d_i \in \mathcal{D}^t$. Chen et al. [2015] used the following objective function for each positive document d_i (a similar objective function can also be formulated for each negative document):

$$F_{+,i} = P(+|d_i) - P(-|d_i) \ . \tag{3.23}$$

We omit the derivation steps and just give the final equations below. To simplify the equations, we define $g(X)$, a function of X where X is a vector consisting of $X_{+,w}$ and $X_{-,w}$ of each word w:

$$g(X) = \beta^{|d_i|} = \left(\frac{\lambda |V| + \sum_{v=1}^{|V|} X_{+,v}}{\lambda |V| + \sum_{v=1}^{|V|} X_{-,v}} \right)^{|d_i|} , \tag{3.24}$$

$$\frac{\partial F_{+,i}}{\partial X_{+,u}} = \frac{\frac{n_{u,d_i}}{\lambda + X_{+,u}} + \frac{P(-)}{P(+)} \prod_{w \in d_i} \left(\frac{\lambda + X_{-,w}}{\lambda + X_{+,w}} \right)^{n_{w,d_i}} \times \frac{\partial g}{\partial X_{+,u}}}{1 + \frac{P(-)}{P(+)} \prod_{w \in d_i} \left(\frac{\lambda + X_{-,w}}{\lambda + X_{+,w}} \right)^{n_{w,d_i}} \times g(X)} - \frac{n_{u,d_i}}{\lambda + X_{+,u}} , \tag{3.25}$$

$$\frac{\partial F_{+,i}}{\partial X_{-,u}} = \frac{\frac{n_{u,d_i}}{\lambda + X_{-,u}} \times g(X) + \frac{\partial g}{\partial X_{-,u}}}{\frac{P(+)}{P(-)} \prod_{w \in d_i} \left(\frac{\lambda + X_{+,w}}{\lambda + X_{-,w}} \right)^{n_{w,d_i}} + g(X)} . \tag{3.26}$$

In stochastic gradient descent, we update the variables $X_{+,u}$ and $X_{-,u}$ for the positive document d_i iteratively using:

$$X_{+,u}^{l} = X_{+,u}^{l-1} - \gamma \frac{\partial F_{+,i}}{\partial X_{+,u}} \ , \ \text{and}$$

$$X_{-,u}^{l} = X_{-,u}^{l-1} - \gamma \frac{\partial F_{+,i}}{\partial X_{-,u}} \ ,$$

where u represents each word in d_i. γ is the learning rate and l represents each iteration. Similar update rules can be derived for each negative document d_i. $X^0_{+,u} = N^t_{+,u} + N^{KB}_{+,u}$ and $X^0_{-,u} = N^t_{-,u} + N^{KB}_{-,u}$ serve as the starting points. The iterative updating process stops when the counts converge.

Exploiting Knowledge via Penalty Terms

The above optimization can update the virtual counts for better classification in the current domain. However, it does not deal with the issue of domain-dependent sentiment words, i.e., some words may change their polarities across different domains. Nor does it use the domain-level knowledge in the knowledge base *KB* (Section 3.5.2). We thus propose to add penalty terms into the optimization to accomplish these.

The idea is that if a word w can distinguish classes very well in the current domain training data, we should rely more on the current domain training data. So we define a set V_T of distinguishing words in the current domain. A word w belongs to V_T if $P(w|+)$ is much larger or much smaller than $P(w|-)$ in the current domain, i.e., $\frac{P(w|+)}{P(w|-)} \geq \sigma$ or $\frac{P(w|-)}{P(w|+)} \geq \sigma$, where σ is a parameter. Such words are already effective in classification for the current domain, so the virtual counts in optimization should follow the empirical counts ($N^t_{+,w}$ and $N^t_{-,w}$) in the current task/domain, which are reflected in the L2 regularization penalty term below (α is the regularization coefficient):

$$\frac{1}{2}\alpha \sum_{w \in V_T} \left(\left(X_{+,w} - N^t_{+,w}\right)^2 + \left(X_{-,w} - N^t_{-,w}\right)^2 \right) . \tag{3.27}$$

To leverage domain-level knowledge (the second type of knowledge in the *KB* in Section 3.5.2, we want to use only those reliable parts of the knowledge. The rationale here is that if a word only appears in one or two past domains, the knowledge associated with it is probably not reliable or it is highly specific to those domains. Based on this idea, domain frequency is used to define the reliability of the domain-level knowledge. For w, if $M^{KB}_{+,w} \geq \tau$ or $M^{KB}_{-,w} \geq \tau$ (τ is a parameter), it is regarded as appearing in a reasonable number of domains, making its knowledge reliable. The set of such words is denoted by V_S. Then the second penalty term is:

$$\frac{1}{2}\alpha \sum_{w \in V_S} \left(X_{+,w} - R_w \times X^0_{+,w}\right)^2 + \frac{1}{2}\alpha \sum_{w \in V_S} \left(X_{-,w} - (1 - R_w) \times X^0_{-,w}\right)^2 , \tag{3.28}$$

where the ratio R_w is defined as $M^{KB}_{+,w}/(M^{KB}_{+,w} + M^{KB}_{-,w})$. $X^0_{+,w}$ and $X^0_{-,w}$ are the starting points for SGD. Finally, the partial derivatives in Equations (3.24), (3.25), and (3.26) are revised by adding the corresponding partial derivatives of Equations (3.27) and (3.28) to them.

3.5.4   DISCUSSIONS

Here we want to discuss a possible improvement to LSC, and a related lifelong sentiment classification work based on voting.

Possible Improvements to LSC

So far we have discussed how to improve the future task learning by leveraging the prior probability knowledge gained from learning the past tasks. One question is whether we can also use the future learning results to go back to help past learning. This is possible because we can apply the same LSC technique by treating the past task to be improved as the future task and the rest of all tasks as the past tasks. The weakness of this approach is that we need the training data of the past task. But what happens if the past task training data is forgotten (like that in human learning)? This is an interesting research problem, and I believe it is possible.

Lifelong Sentiment Classification via Voting

Xia et al. [2017] presented two LL methods for sentiment classification via voting of individual task classifiers. The first method votes with equal weight for each task classifier. This method can be applied to help past tasks. The second method uses weighted voting. However, like LSC, it needs the past task training data to improve its model. Furthermore, their tasks are actually from the same domain as they partitioned the same dataset into subsets and treated each subset as a task. Tasks in LSC are from different domains (different types of products).

3.6   DOMAIN WORD EMBEDDING VIA META-LEARNING

LL can also be realized through meta-learning. This section describes such a method, which aims to improve word embeddings for a domain without a large corpus. Learning word embeddings [Mikolov et al., 2013a,b, Mnih and Hinton, 2007, Pennington et al., 2014, Turian et al., 2010] has received a significant amount of attention in recent years due to its success in numerous natural language processing (NLP) applications. The "secret sauce" of the success of word embeddings is that a large-scale corpus can be turned into a huge number (e.g., billions) of training examples to learn the "semantic meanings" of words, which are used to perform many down-stream NLP tasks. Two implicit assumptions are often made about the effectiveness of embeddings to down-stream tasks: (1) the training corpus for embedding is available and much larger than the training data of the down-stream task, and (2) the topic (domain) of the embedding corpus is closely aligned with the topic of the down-stream NLP task. However, many real-world applications do not meet both assumptions.

In most cases, the in-domain corpus is of limited size, which is insufficient for training good embeddings. In such applications, researchers and practitioners often just use some general-purpose embeddings that are trained using a very large general-purpose corpus (which satisfies the first assumption) covering almost all possible topics, e.g., the well-known GloVe embeddings [Pennington et al., 2014] trained using 840 billion tokens covering almost all topics or domains on the Web. Such embeddings have been shown to work reasonably well in many domain-specific tasks. This is not surprising as the meaning of a word is largely shared across domains and tasks. However, this solution violates the second assumption, which often leads to sub-optimal results for domain-specific tasks [Xu et al., 2018]. One obvious explanation for this is that the general-purpose embeddings do provide some useful information for many words in the domain task, but their embedding representations may not be ideal for the domain and in some cases they may even conflict with the meanings of some words in the task domain because words often have multiple senses or meanings. For instance, we have a task in the programming domain, which has the word "Java." A large-scale general-purpose corpus, which is very likely to include texts about coffee shops, supermarkets, the Java island of Indonesia, etc., can easily squeeze the room for representing the "Java" context words like "function," "variable" and "Python" in the programming domain. This results in a poor representation of the word "Java" for the programming domain task.

To solve this problem and also the limited in-domain corpus size problem, cross-domain embeddings have been investigated [Bollegala et al., 2015, 2017, Yang et al., 2017] via transfer learning. These methods allow some in-domain words to leverage the general-purpose embeddings in the hope that the meanings of these words in the general-purpose embeddings do not deviate too much from the in-domain meanings of these words. The embeddings of these words can thus be improved. However, these methods cannot improve the embeddings of many other words with domain-specific meanings (e.g., "Java"). Furthermore, some words in the general-purpose embeddings may carry meanings that are different than those in the task domain.

Xu et al. [2018] proposed to improve domain embedding via LL by expanding the in-domain corpus. The problem is stated as follows: Assuming that the learning system has seen corpora of N domains in the past: $D_{1:N} = \{D_1, \ldots, D_N\}$, when a new task arrives with a domain corpus D_{N+1}, the system automatically generates word embeddings for the $(N+1)$-th domain by leveraging some useful information or knowledge from the past N domains.

The main challenges of this problem are twofold: (1) how to automatically identify relevant information/knowledge from the past N domains without the help of a human user and (2) how to integrate the relevant information into the new $(N+1)$-th domain corpus. Xu et al. [2018] proposed a meta-learning based algorithm called L-DEM (Lifelong Domain Embedding via Meta-learning) to tackle the challenges.

To deal with the first challenge, for a word in the new domain, L-DEM learns to identify similar contexts of the word in the past domains. Here the context of a word in a domain means the surrounding words of that word in the domain corpus, called the *domain context* of the word. For this, a multi-domain meta-learner is introduced. The meta-learner learns a meta-predictor using data (corpora) from multiple domains. This meta-predictor is called the base predictor. When a new domain arrives with its corpus, the system first adapts the base meta-predictor to make it suitable for the new domain. The resulting domain-specific meta-predictor is then used to identify similar (or relevant) domain contexts in each of the past domain for each word in the new domain. The training data for meta-learning and domain adaptation are produced automatically. To tackle the second challenge, L-DEM augments the new domain corpus with the relevant domain contexts (knowledge) produced by the meta-predictor from the past domain corpora and uses the combined data to train the embeddings for the new domain.

For example, for the word "Java" in the programming domain (the new domain), the meta-predictor may produce similar domain contexts from some previous domains like programming language, software engineering, operating systems, etc. These domain contexts will be combined with the new domain corpus for "Java" to train a new domain embedding for "Java." The detailed technique is involved. Interested readers, please refer to Xu et al. [2018].

3.7   SUMMARY AND EVALUATION DATASETS

Although LL started with supervised learning more than 20 years ago, existing work is still limited in both variety and in depth. There is still no general mechanism or algorithm that can be applied to any sequence of tasks like existing ML algorithms such as SVM, Naïve Bayes, or deep learning, which can be used for almost any supervised learning task. There are many reasons for this. Perhaps, the most important reason is that the research community still does not have a good understanding of what the knowledge is in general, how to represent knowledge, and how to use knowledge in learning effectively. A unified theory of knowledge and the related issues is urgently needed. Another reason is that knowledge from supervised learning is difficult to use across domains because to some extent optimization is an obstacle for reuse or transfer because each model is highly optimized for its specific task. It is difficult to pick and choose some pieces of knowledge learned from previous tasks or domains and apply them to the new tasks because a model is often not decomposable. For example, it is very difficult to reuse any knowledge in an SVM model or apply it in different but similar tasks. Simpler models are often much easier to reuse. For example, it is not hard to pick some rules from a rule-based classifier and use them to help learn a new task. This is probably why human learning is not optimized as the human brains are not good at optimization and also our intelligence requires flexibility.

Evaluation Datasets: To help researchers working in the field, we summarize the evaluation datasets used in the papers covered in the chapter. For those publicly available datasets, we provide their URLs.

Thrun [1996b] used a dataset of color camera images of different objects (such as bottle, hammer, and book) in the evaluation. Caruana [1997] used the dataset 1D-ALVINN [Pomerleau, 2012] in the road-following domain. They also created the dataset 1D-DOORS [Caruana, 1997] in the object-recognition domain. In addition, a medical decision-making application was also tested in Caruana [1997]. Ruvolo and Eaton [2013b] used three datasets in their evaluation. The first is the land mine dataset from Xue et al. [2007], which detects whether or not a land mine appears in an area according to radar images. The second is the facial expression recognition challenge dataset in Valstar et al. [2011].[1] The third is a London Schools dataset.[2] Chen et al. [2015] evaluated using Amazon reviews from 20 diverse product domains, which is a subset of the dataset in Chen and Liu [2014b].[3] Xu et al. [2018] used the Amazon Review datasets

[1]http://gemep-db.sspnet.eu
[2]https://github.com/tjanez/PyMTL/tree/master/data/school
[3]https://www.cs.uic.edu/~zchen/downloads/KDD2014-Chen-Dataset.zip

from He and McAuley [2016], which is a collection of multiple-domain corpora organized in multiple levels. The paper considered each second-level category (the first level is department) as a domain and aggregate all reviews under each category as one domain corpus. This ends up with a very diverse domain collection.

CHAPTER 4

Continual Learning and Catastrophic Forgetting

In the recent years, lifelong learning (LL) has attracted a great deal of attention in the deep learning community, where it is often called *continual learning*. Though it is well-known that deep neural networks (DNNs) have achieved state-of-the-art performances in many machine learning (ML) tasks, the standard multi-layer perceptron (MLP) architecture and DNNs suffer from *catastrophic forgetting* [McCloskey and Cohen, 1989] which makes it difficult for continual learning. The problem is that when a neural network is used to learn a sequence of tasks, the learning of the later tasks may degrade the performance of the models learned for the earlier tasks. Our human brains, however, seem to have this remarkable ability to learn a large number of different tasks without any of them negatively interfering with each other. Continual learning algorithms try to achieve this same ability for the neural networks and to solve the catastrophic forgetting problem. Thus, in essence, continual learning performs incremental learning of new tasks. Unlike many other LL techniques, the emphasis of current continual learning algorithms has not been on how to leverage the knowledge learned in previous tasks to help learn the new task better. In this chapter, we first give an overview of catastrophic forgetting (Section 4.1) and survey the proposed continual learning techniques that address the problem (Section 4.2). We then introduce several recent continual learning methods in more detail (Sections 4.3–4.8). Two evaluation papers are also covered in Section 4.9 to evaluate the performances of some existing continual learning algorithms. Last but not least, we give a summary of the chapter and list the relevant evaluation datasets.

4.1 CATASTROPHIC FORGETTING

Catastrophic forgetting or *catastrophic interference* was first recognized by McCloskey and Cohen [1989]. They found that, when training on new tasks or categories, a neural network tends to forget the information learned in the previous trained tasks. This usually means a new task will likely override the weights that have been learned in the past, and thus degrade the model performance for the past tasks. Without fixing this problem, a single neural network will not be able to adapt itself to an LL scenario, because it *forgets* the existing information/knowledge when it learns new things. This was also referred to as the stability-plasticity dilemma in Abraham and Robins [2005]. On the one hand, if a model is too stable, it will not be able to consume new information from the future training data. On the other hand, a model with sufficient plasticity suffers from large weight changes and forgets previously learned representations. We should note that catastrophic forgetting happens to traditional multi-layer perceptrons as well as to DNNs. Shadow single-layered models, such as self-organizing feature maps, have been shown to have catastrophic interference too [Richardson and Thomas, 2008].

A concrete example of catastrophic forgetting is transfer learning using a deep neural network. In a typical transfer learning setting, where the source domain has plenty of labeled data and the target domain has little labeled data, *fine-tuning* is widely used in DNNs [Dauphin et al., 2012] to adapt the model for the source domain to the target domain. Before fine-tuning, the source domain labeled data is used to pre-train the neural network. Then the output layers of this neural network are retrained given the target domain data. Backpropagation-based fine-tuning is applied to adapt the source model to the target domain. However, such an approach suffers from catastrophic forgetting because the adaptation to the target domain usually disrupts the weights learned for the source domain, resulting inferior inference in the source domain.

Li and Hoiem [2016] presented an excellent overview of the traditional methods for dealing with catastrophic forgetting. They characterized three sets of parameters in a typical approach:

- θ_s: set of parameters shared across all tasks;

- θ_o: set of parameters learned specifically for previous tasks; and

- θ_n: randomly initialized task-specific parameters for new tasks.

Li and Hoiem [2016] gave an example in the context of image classification, in which θ_s consists of five convolutional layers and two fully connected layers in the AlexNet architecture [Krizhevsky et al., 2012]. θ_o is the output layer for classification [Russakovsky et al., 2015] and its corresponding weights. θ_n is the output layer for new tasks, e.g., scene classifiers.

There are three traditional approaches to learning θ_n with knowledge transferred from θ_s.

- **Feature Extraction** (e.g., Donahue et al. [2014]): both θ_s and θ_o remain the same while the outputs of some layers are used as features for training θ_n for the new task.

- **Fine-tuning** (e.g., Dauphin et al. [2012]): θ_s and θ_n are optimized and updated for the new task while θ_o remains fixed. To prevent large shift in θ_s, a low learning rate is typically applied. Also, for the similar purpose, the network can be *duplicated and fine-tuned* for each new task, leading to N networks for N tasks. Another variation is to fine-tune parts of θ_s, for example, the top layers. This can be seen as a compromise between fine-tuning and feature extraction.

- **Joint Training** (e.g., Caruana [1997]): All the parameters θ_s, θ_o, θ_n are jointly optimized across all tasks. This requires storing all the training data of all tasks. Multi-task learning (MTL) typically takes this approach.

The pros and cons of these methods are summarized in Table 4.1. In light of these pros and cons, Li and Hoiem [2016] proposed an algorithm called "Learning without Forgetting" that explicitly deals with the weaknesses of these methods; see Section 4.3.

Table 4.1: Summary of traditional methods for dealing with catastrophic forgetting. Adapted from Li and Hoiem [2016].

Category	Feature Extraction	Fine-Tuning	Duplicate and Fine-Tuning	Joint Training
New task performance	Medium	Good	Good	Good
Old task performance	Good	Bad	Good	Good
Training efficiency	Fast	Fast	Fast	Slow
Testing efficiency	Fast	Fast	Slow	Fast
Storage requirement	Medium	Medium	Large	Large
Require previous task data	No	No	No	Yes

4.2   CONTINUAL LEARNING IN NEURAL NETWORKS

A number of continual learning approaches have been proposed to lessen catastrophic forgetting recently. This section gives an overview for these newer developments. A comprehensive survey on the same topic is also given in Parisi et al. [2018a].

Much of the existing work focuses on *supervised learning* [Parisi et al., 2018a]. Inspired by fine-tuning, Rusu et al. [2016] proposed a progressive neural network that retains a pool of pretrained models and learns lateral connections among them. Kirkpatrick et al. [2017] proposed a model called Elastic Weight Consolidation (EWC) that quantifies the importance of weights to previous tasks, and selectively adjusts the plasticity of weights. Rebuffi et al. [2017] tackled the LL problem by retaining an exemplar set that best approximates the previous tasks. A network of experts is proposed by Aljundi et al. [2016] to measure task relatedness for dealing with catastrophic forgetting. Rannen Ep Triki et al. [2017] used the idea of autoencoder to extend the method in "Learning without Forgetting" [Li and Hoiem, 2016]. Shin et al. [2017] followed the Generative Adversarial Networks (GANs) framework [Goodfellow, 2016] to keep a set of generators for previous tasks, and then learn parameters that fit a mixed set of real data of the new task and replayed data of previous tasks. All these works will be covered in details in the next few sections.

Instead of using knowledge distillation as in the model "Learning without Forgetting" (LwF) [Li and Hoiem, 2016], Jung et al. [2016] proposed a less-forgetful learning that regularizes the final hidden activations. Rosenfeld and Tsotsos [2017] proposed controller modules to optimize loss on the new task with representations learned from previous tasks. They found that they could achieve satisfactory performance while only requiring about 22% of parameters of the fine-tuning method. Ans et al. [2004] designed a dual-network architecture to generate pseudo-items which are used to self-refresh the previous tasks. Jin and Sendhoff [2006] modeled the catastrophic forgetting problem as a multi-objective learning problem and proposed a multi-objective pseudo-rehearsal framework to interleave base patterns with new patterns during optimization. Nguyen et al. [2017] proposed variational continual learning by combining online variational inference (VI) and Monte Carlo VI for neural networks. Motivated by EWC [Kirkpatrick et al., 2017], Zenke et al. [2017] measured the synapse consolidation strength in an online fashion and used it as regularization in neural networks. Seff et al. [2017] proposed to solve continual generative modeling by combining the ideas of GANs [Goodfellow, 2016] and EWC [Kirkpatrick et al., 2017].

Apart from regularization-based approaches mentioned above (e.g., LwF [Li and Hoiem, 2016], EWC [Kirkpatrick et al., 2017]), dual-memory-based learning systems have also been proposed for LL. They are inspired by the complementary learning systems (CLS) theory [Kumaran et al., 2016, McClelland et al., 1995] in which memory consolidation and retrieval are related to the interplay of the mammalian hippocampus (short-term memory) and neocortex (long-term memory). Gepperth and Karaoguz [2016] proposed using a modified self-organizing map (SOM) as the long-term memory. To complement it, a short-term memory (STM) is added to store novel examples. During the sleep phase, the whole content of STM is replayed to the system. This process is known as intrinsic replay or pseudo-rehearsal [Robins, 1995]. It trains all the nodes in the network with new data (e.g., from STM) and replayed samples from previously seen classes or distributions on which the network has been trained. The replayed samples prevents the network from forgetting. Kemker and Kanan [2018] proposed a similar dual-memory system called FearNet. It uses a hippocampal network for STM, a medial prefrontal cortex (mPFC) network for long-term memory, and a third neural network to determine which memory to use for prediction. More recent developments in this direction include Deep Generative Replay [Shin et al., 2017], DGDMN [Kamra et al., 2017] and Dual-Memory Recurrent Self-Organization [Parisi et al., 2018b].

Some other related works include Learn++ [Polikar et al., 2001], Gradient Episodic Memory [Lopez-Paz et al., 2017], Pathnet [Fernando et al., 2017], Memory Aware Synapses [Aljundi et al., 2017], One Big Net for Everything [Schmidhuber, 2018], Phantom Sampling [Venkatesan et al., 2017], Active Long Term Memory Networks [Furlanello et al., 2016], Conceptor-Aided Backprop [He and Jaeger, 2018], Gating Networks [Masse et al., 2018, Serrà et al., 2018], PackNet [Mallya and Lazebnik, 2017], Diffusion-based Neuromodulation [Velez and Clune, 2017], Incremental Moment Matching [Lee et al., 2017b], Dynamically Expandable Networks [Lee et al., 2017a], and Incremental Regularized Least Squares [Camoriano et al., 2017].

There are some *unsupervised learning* works as well. Goodrich and Arel [2014] studied unsupervised online clustering in neural networks to help mitigate catastrophic forgetting. They proposed building a path through the neural network to select neurons during the feed-forward pass. Each neural is assigned with a cluster centroid, in addition to the regular weights. In the new task, when a sample arrives, only the neurons whose cluster centroid points are close to the sample are selected. This can be viewed as a special dropout training [Hinton et al., 2012]. Parisi et al. [2017] tackled LL of action representations by learning unsupervised visual representation. Such representations are incrementally associated with action labels based on occurrence frequency. The proposed model achieves competitive performance compared to models trained with predefined number of action classes.

In the *reinforcement learning* applications [Ring, 1994], other than the works mentioned above (e.g., Kirkpatrick et al. [2017], Rusu et al. [2016]), Mankowitz et al. [2018] proposed a continual learning agent architecture called Unicorn. The Unicorn agent is designed to have the ability to simultaneously learn about multiple tasks including the new tasks. The agent can reuse its accumulated knowledge to solve related tasks effectively. Last but not least, the architecture aims to aid agent in solving tasks with deep dependencies. The essential idea is to learn multiple tasks off-policy, i.e., when acting on-policy with respect to one task, it can use this experience to update policies of related tasks. Kaplanis et al. [2018] took the inspiration from biological synapses and incorporated different timescales of plasticity to mitigate catastrophic forgetting over multiple timescales. Its idea of synaptic consolidation is along the lines of EWC [Kirkpatrick et al., 2017]. Lipton et al. [2016] proposed a new reward shaping function that learns the probability of imminent catastrophes. They named it as *intrinsic fear*, which is used to penalize the Q-learning objective.

Evaluation frameworks were also proposed in the context of catastrophic forgetting. Goodfellow et al. [2013a] evaluated traditional approaches including dropout training [Hinton et al., 2012] and various activation functions. More recent continual learning models were evaluated in Kemker et al. [2018]. Kemker et al. [2018] used large-scale datasets and evaluated model accuracy on both old and new tasks in the LL setting. See Section 4.9 for more details. In the next few sections, we discuss some representative continual learning approaches.

4.3 LEARNING WITHOUT FORGETTING

This section describes the approach called *Learning without Forgetting* given in Li and Hoiem [2016]. Based on the notations in Section 4.1, it learns θ_n (parameters for the new task) with the help of θ_s (shared parameters for all tasks) and θ_o (parameters for old tasks) without degrading much of the performance on the old tasks. The idea is to optimize θ_s and θ_n on the new task with the constraint that the predictions on the new task's examples using θ_s and θ_o do not shift much. The constraint makes sure that the model still "remembers" its old parameters, for the sake of maintaining satisfactory performance on the previous tasks.

The algorithm is outlined in Algorithm 4.1. Line 2 records the predictions Y_o of the new task's examples X_n using θ_s and θ_o, which will be used in the objective function (Line 7). For each new task, nodes are added to the output layer, which is fully connected to the layer beneath.

These new nodes are first initialized with random weights θ_n (Line 3). There are three parts in the objective function in Line 7.

Algorithm 4.1 Learning without Forgetting

Input: shared parameters θ_s, task-specific parameters for old tasks θ_o, training data X_n, Y_n for the new task.
Output: updated parameters θ_s^*, θ_o^*, θ_n^*.

1: // Initialization phase.
2: $Y_o \leftarrow \text{CNN}(X_n, \theta_s, \theta_o)$
3: $\theta_n \leftarrow \text{RANDINIT}(|\theta_n|)$
4: // Training phase.
5: Define $\hat{Y}_n \equiv \text{CNN}(X_n, \hat{\theta}_s, \hat{\theta}_n)$
6: Define $\hat{Y}_o \equiv \text{CNN}(X_n, \hat{\theta}_s, \hat{\theta}_o)$
7: $\theta_s^*, \theta_o^*, \theta_n^* \leftarrow \text{argmin}_{\hat{\theta}_s, \hat{\theta}_o, \hat{\theta}_n} \left(\mathcal{L}_{new}(\hat{Y}_n, Y_n) + \lambda_o \mathcal{L}_{old}(\hat{Y}_o, Y_o) + \mathcal{R}(\theta_s, \theta_o, \theta_n) \right)$

- $\mathcal{L}_{new}(\hat{Y}_n, Y_n)$: minimize the difference between the predicted values $\hat{Y}_n$ and the groundtruth Y_n. $\hat{Y}_n$ is the predicted value using the current parameters $\hat{\theta}_s$ and $\hat{\theta}_n$ (Line 5). In Li and Hoiem [2016], the multinomial logistic loss is used:

$$\mathcal{L}_{new}(\hat{Y}_n, Y_n) = -Y_n \cdot \log \hat{Y}_n .$$

- $\mathcal{L}_{old}(\hat{Y}_o, Y_o)$: minimize the difference between the predicted values $\hat{Y}_o$ and the recorded values Y_o (Line 2), where $\hat{Y}_o$ comes from the current parameters $\hat{\theta}_s$ and $\hat{\theta}_o$ (Line 6). Li and Hoiem [2016] used knowledge distillation loss [Hinton et al., 2015] to encourage the outputs of one network to approximate the outputs of another. The distillation loss is defined as modified cross-entropy loss:

$$\mathcal{L}_{old}(\hat{Y}_o, Y_o) = -H(\hat{Y}_o', Y_o')$$
$$= -\sum_{i=1}^{l} y_o'^{(i)} \log \hat{y}_o'^{(i)} ,$$

where l is the number of labels. $y_o'^{(i)}$ and $\hat{y}_o'^{(i)}$ are the modified probabilities defined as:

$$y_o'^{(i)} = \frac{(y_o^{(i)})^{1/T}}{\sum_j (y_o^{(j)})^{1/T}}, \quad \hat{y}_o'^{(i)} = \frac{(\hat{y}_o^{(i)})^{1/T}}{\sum_j (\hat{y}_o^{(j)})^{1/T}} .$$

T is set to 2 in Li and Hoiem [2016] to increase the weights of smaller logit values. In the objective function (Line 7), λ_o is used to balance the new task and the old/past tasks. Li and Hoiem [2016] tried various values for λ_o in their experiments.

- $\mathcal{R}(\theta_s, \theta_o, \theta_n)$: regularization term to avoid overfitting.

4.4 PROGRESSIVE NEURAL NETWORKS

Progressive neural networks were proposed by Rusu et al. [2016] to explicitly tackle catastrophic forgetting for the problem of LL. The idea is to keep a pool of pretrained models as knowledge, and use lateral connections between them to adapt to the new task. The model was originally proposed to tackle reinforcement learning, but the model architecture is general enough for other ML paradigms such as supervised learning. Assuming there are N existing/past tasks: $\mathcal{T}_1, \mathcal{T}_2, \ldots,$ $\mathcal{T}_N$, progressive neural networks maintain N neural networks (or N columns). When a new task $\mathcal{T}_{N+1}$ is created, a new neural network (or a new column) is created, and its lateral connections with all previous tasks are learned. The mathematical formulation is presented below.

In progressive neural networks, each task $\mathcal{T}_n$ is associated with a neural network, which is assumed to have L layers with hidden activations $h_i^{(n)}$ for the units at layer $i \leq L$. The set of parameters in the neural network for $\mathcal{T}_n$ is denoted by $\Theta^{(n)}$. When a new task $\mathcal{T}_{N+1}$ arrives, the parameters $\Theta^{(1)}, \Theta^{(2)}, \ldots, \Theta^{(N)}$ remain the same while each layer $h_i^{(N+1)}$ in the $\mathcal{T}_{N+1}$'s neural network takes inputs from $(i-1)$th layers of all previous tasks' neural networks, i.e.,

$$h_i^{(N+1)} = \max \left(0, W_i^{(N+1)} h_{i-1}^{(N+1)} + \sum_{n < N+1} U_i^{(n:N+1)} h_{i-1}^{(n)} \right) , \qquad (4.1)$$

where $W_i^{(N+1)}$ denotes the weight matrix of layer i in neural network $N+1$. The lateral connections are learned via $U_i^{(n:N+1)}$ to indicate how strong the $(i-1)$th layer from task n influences the ith layer from task $N+1$. h_0 is the network input.

Unlike pretraining and fine-tuning, progressive neural networks do not assume any relationship between tasks, which makes it more practical for real-world applications. The lateral connections can be learned for related, orthogonal, or even adversarial tasks. To avoid catastrophic forgetting, settings of parameters $\Theta^{(n)}$ for existing tasks $\mathcal{T}_n$ where $n \leq N$ are "frozen" while the new parameter set $\Theta^{(N+1)}$ is learned and adapted for the new task $\mathcal{T}_{N+1}$. As a result, the performance on existing tasks does not degrade.

For the applications in reinforcement learning, each task's neural network is trained to learn a policy function for a particular Markov Decision Process (MDP). The policy function implies the probabilities over actions given states. Nonlinear lateral connections are learned through a single hidden perceptron layer, which reduces the number of parameters from the lateral connections to the same order as $|\Theta^{(1)}|$. More details can be found in Rusu et al. [2016].

With the flexibility of considering various task relationships, progressive neural networks come at a price: it can explode the numbers of parameters with an increasing number of tasks, since it needs to learn a new neural network for a new task and its lateral connections with all existing ones. Rusu et al. [2016] suggested pruning [LeCun et al., 1990] or online compression [Rusu et al., 2015] as potential solutions.

4.5   ELASTIC WEIGHT CONSOLIDATION

Kirkpatrick et al. [2017] proposed a model called *Elastic Weight Consolidation* (EWC) to mitigate catastrophic forgetting in neural networks. It was inspired by human brain in which synaptic consolidation enables continual learning by reducing the plasticity of synapses related to previous learned tasks. As mentioned in Section 4.1, plasticity is the main cause of catastrophic forgetting since the weights learned in the previous tasks can be easily modified given a new task. More precisely, plasticity of weights that are closely related to previous tasks is more prone to catastrophic forgetting than plasticity of weights that are loosely connected to previous tasks. This motivates [Kirkpatrick et al., 2017] to quantify the importance of weights in terms of their impact on previous tasks' performance, and selectively decrease the plasticity of those important weights to previous tasks.

Kirkpatrick et al. [2017] illustrated their idea using an example consisting of two tasks A and B where A is a previous task and B is the new task. The example only contains two tasks for easy understanding, but the EWC model works in an LL manner with tasks coming in a sequence. The parameters (weights and biases) for task A and B are represented by θ_A and θ_B. The sets of parameters that lead to low errors for task A and B are represented by Θ_A^* and Θ_B^*, respectively. Over-parametrization makes it possible to find a solution $\theta_B^* \in \Theta_B^*$ and $\theta_B^* \in \Theta_A^*$, i.e., the solution is learned toward task B while also maintaining low errors in task A. EWC achieves this goal by constraining the parameters to stay in A's low-error region. Figure 4.1 visualizes the example.

The Bayesian approach is used to measure the importance of parameters toward a task in EWC. In particular, the importance is modeled as the posterior distribution $p(\theta|\mathcal{D})$, the probability of parameter θ given a task's training data $\mathcal{D}$. Using Bayes' rule, the log value of the posterior probability is:

$$\log p(\theta|\mathcal{D}) = \log p(\mathcal{D}|\theta) + \log p(\theta) - \log p(\mathcal{D}) \ . \tag{4.2}$$

Assume that the data consists of two independent parts: $\mathcal{D}_A$ for task A and $\mathcal{D}_B$ for task B. Equation (4.2) can be written as:

$$\log p(\theta|\mathcal{D}) = \log p(\mathcal{D}_B|\theta) + \log p(\theta|\mathcal{D}_A) - \log p(\mathcal{D}_B) \ . \tag{4.3}$$

The left side in Equation (4.3) is still the posterior distribution given the *entire* dataset, while the right side only depends on the loss function for task B, i.e., $\log p(\mathcal{D}_B|\theta)$. All the information related to task A is embedded in the term $\log p(\theta|\mathcal{D}_A)$. EWC wants to extract information about weight importance from $\log p(\theta|\mathcal{D}_A)$. Unfortunately, $\log p(\theta|\mathcal{D}_A)$ is intractable. Thus, EWC approximates it as a Gaussian distribution with mean given by the parameters θ_A^* and a diagonal precision by the diagonal of the Fisher information matrix F. Thus, the new loss function in EWC is:

$$\mathcal{L}(\theta) = \mathcal{L}_B(\theta) + \sum_i \frac{\lambda}{2} F_i (\theta_i - \theta_{A,i}^*)^2 \ , \tag{4.4}$$

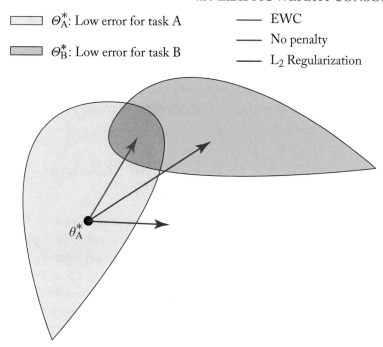

Figure 4.1: An example to illustrate EWC. Given task B, a regular neural network learns a point that yields a low error for task B but not task A (blue arrow). A L_2 regularization instead provides a suboptimal model to task B (purple arrow). EWC updates its parameters for task B while slowly updating the parameters important to task A to stay in A's low error region (red arrow).

where $\mathcal{L}_B(\theta)$ is the loss for task B only. λ controls how strong the constraint posed should not move too far away from task A's low error area. i denotes each index in the weight vector.

Recall that if θ has n dimensions, $\theta_1, \theta_2, \ldots, \theta_n$, the Fisher information matrix F is a $n \times n$ matrix with each entry being:

$$I(\theta)_{ij} = E_X\left[\left(\frac{\partial}{\partial\theta_i}\log p(\mathcal{D}|\theta)\right)\left(\frac{\partial}{\partial\theta_j}\log p(\mathcal{D}|\theta)\right)\bigg|\theta\right]. \tag{4.5}$$

The diagonal entry is then:

$$F_i = I(\theta)_{ii} = E_X\left[\left(\frac{\partial}{\partial\theta_i}\log p(\mathcal{D}|\theta)\right)^2\bigg|\theta\right]. \tag{4.6}$$

When a task C comes, EWC updates Equation (4.4) with the penalty term enforcing the parameters θ to be close to $\theta_{A,B}^*$, where $\theta_{A,B}^*$ is the parameters learned for tasks A and B.

To evaluate EWC, Kirkpatrick et al. [2017] used the MNIST dataset [LeCun et al., 1998]. A new task is obtained by generating a random permutation and the input pixels of all images are shuffled according to the permutation. As a result, each task is unique with equal difficulty to the original MNIST problem. The results showed that EWC achieves superior performances to those models that suffer from catastrophic forgetting. For more details on the evaluation as well as EWC's application in reinforcement learning, please refer to the original paper by Kirkpatrick et al. [2017].

4.6   ICARL: INCREMENTAL CLASSIFIER AND REPRESENTATION LEARNING

Rebuffi et al. [2017] proposed a new model for *class-incremental learning*. Class-incremental learning requires the classification system to incrementally learn and classify new classes that it has never seen before. This is similar to *open-world-learning* (or *cumulative learning*) [Fei et al., 2016] introduced in Chapter 5 without the rejection capability of open-world learning. It assumes that examples of different classes can occur at different times, with which the system should maintain a satisfactory classification performance on each observed class. Rebuffi et al. [2017] also emphasized that computational resources should be bounded or slowly increased with more and more classes coming.

To meet these criteria, a new model called *iCaRL, incremental Classifier and Representation Learning*, was designed to simultaneously learn classifiers and feature representations in the class-incremental setting. At the high level, iCaRL maintains a set of exemplar examples for each observed class. For each class, an exemplar set is a subset of all examples of the class, aiming to carry the most representative information of the class. The classification of a new example is performed by choosing the class whose exemplars are the most similar to it. When a new class shows up, iCaRL creates an exemplar set for this new class while trimming the exemplar sets of the existing/previous classes.

Formally, at any time, iCaRL learns a stream of classes in the class-incremental learning setting with their training example sets, X^s, X^{s+1}, ..., X^t, where X^y is a set of examples of class y. y can either be an observed/past class or a new class. To avoid memory overflow, iCaRL holds a fixed number (K) of exemplars in total. With C classes, the exemplar sets are represented by $\mathcal{P} = \{P_1, \ldots, P_C\}$ where each class's exemplar set P_i maintains K/C exemplars. In Rebuffi et al. [2017], both original examples and exemplars are images, but the proposed method is general enough for non-image datasets.

4.6.1   INCREMENTAL TRAINING

Algorithm 4.2 presents the incremental training algorithm in iCaRL with new training example sets X^s, ..., X^t of classes s, ..., t arriving in a stream. Line 1 updates the model parameters Θ with the new training examples (defined in Algorithm 4.3). Line 2 computes the number of

---

Algorithm 4.2 iCaRL IncrementalTraining

Input: new training examples $X^s, \ldots, X^t$ of new classes $s, \ldots, t$, current model parameters Θ, current exemplar sets $\mathcal{P} = \{P_1, \ldots, P_{s-1}\}$, memory size K.
Output: updated model parameters Θ, updated exemplar sets $\mathcal{P}$.

1: $\Theta \leftarrow$ UpdateRepresentation$(X^s, \ldots, X^t; \mathcal{P}, \Theta)$
2: $m \leftarrow K/t$
3: **for** $y = 1$ **to** $s - 1$ **do**
4:      $P_y \leftarrow P_y[1 : m]$
5: **end for**
6: **for** $y = s$ **to** t **do**
7:      $P_y \leftarrow$ ConstructExemplarSet(X^y, m, Θ)
8: **end for**
9: $\mathcal{P} \leftarrow \{P_1, \ldots, P_t\}$

---

exemplars per class. For each existing class, we reduce the number of exemplars per class to m. Since the exemplars are created in the order of importance (see Algorithm 4.4), we just keep the first m exemplars for each class (Line 3–5). Line 6–8 construct the exemplar set for each new class (see Algorithm 4.4).

4.6.2   UPDATING REPRESENTATION

Algorithm 4.3 details the steps for updating the feature representation. Two datasets are created (Lines 1 and 2): one with all existing exemplar examples, and the other with new examples of the new classes. Note that the exemplar examples have the original feature space, not the learned representation. Lines 3–5 store the prediction output of each exemplar example with the current model. Learning in Rebuffi et al. [2017] used a convolutional neural network (CNN) [LeCun et al., 1998], interpreted as a trainable feature extractor: $\varphi : \mathcal{X} \to \mathbb{R}^d$. A single classification layer is added with as many sigmoid output nodes as the number of classes observed so far. The output score for class $y \in \{1, \ldots, t\}$ is formulated as follows:

$$g_y(x) = \frac{1}{1 + exp(-a_y(x))} \qquad \text{with } a_y(x) = w_y^{\mathrm{T}} \varphi(x) \ . \tag{4.7}$$

Note that the network is just utilized for representation learning, not for the actual classification. The actual classification is covered in Section 4.6.4. The last step in Algorithm 4.3 runs Backpropagation with the loss function that (1) minimizes the loss on the new examples of new classes D^{new} (*classification loss*), and (2) reproduces the scores stored using previous networks (*distillation loss* [Hinton et al., 2015]). The hope is that the neural network will be updated with new examples of the new classes, while not forgetting the existing classes.

Algorithm 4.3 iCaRL UpdateRepresentation

Input: new training examples $X^s, \ldots, X^t$ of new classes $s, \ldots, t$, current model parameters Θ, current exemplar sets $\mathcal{P} = \{P_1, \ldots, P_{s-1}\}$, memory size K.
Output: updated model parameters Θ.

1: $\mathcal{D}^{exemplar} \leftarrow \bigcup_{y=1,\ldots,s-1} \{(x, y) : x \in P_y\}$

2: $\mathcal{D}^{new} \leftarrow \bigcup_{y=s,\ldots,t} \{(x, y) : x \in X^y\}$

3: **for** $y = 1$ **to** $s - 1$ **do**

4:     $q_i^y \leftarrow g_y(x_i)$ for all $(x_i, \cdot) \in \mathcal{D}^{exemplar}$

5: **end for**

6: $\mathcal{D}^{train} \leftarrow \mathcal{D}^{exemplar} \cup \mathcal{D}^{new}$

7: Run network training (e.g., Backpropagation) with loss function that contains *classification* and *distillation* terms:
$$\mathcal{L}(\Theta) = -\sum_{(x_i, y_i) \in \mathcal{D}^{train}} [\sum_{y=s}^{t} \delta_{y=y_i} \log g_y(x_i) + \delta_{y \neq y_i} \log(1 - g_y(x_i)) \\ + \sum_{y=1}^{s-1} q_i^y \log g_y(x_i) + (1 - q_i^y) \log(1 - g_y(x_i))]$$

4.6.3   CONSTRUCTING EXEMPLAR SETS FOR NEW CLASSES

When examples of a new class t show up, iCaRL balances the number of exemplars in each class, i.e., reducing the number of exemplars for each existing class and creating the exemplar set for the new class. If K exemplars are allowed in total due to the memory limitation, each class receives $m = K/t$ exemplar quota. For each existing class, the first m exemplars are kept (Lines 3–5 in Algorithm 4.2). For the new class t, Algorithm 4.4 chooses m exemplars for it. Here is the intuition of how the selection of exemplars works: the average feature vector over all exemplars should be close to the average feature vector over all examples of the class. As such, the general property of all examples in a class does not diminish much when most of them are removed, i.e., only exemplars are retained. Also, to make sure that exemplars can be easily trimmed, the exemplars are stored in the order that the most important ones are stored first, thus making the list a priority list.

In Algorithm 4.4, the average feature vector μ of all training examples of class t is computed (Line 1). Then m exemplars are selected in the order that by picking each exemplar p_k, the average feature vector is the closest to μ compared to adding any other non-exemplar example (Lines 2–4). Consequently, the resulting exemplar set $P \leftarrow (p_1, \ldots, p_m)$ should well approximate the class mean vector. Note that all non-exemplar examples are dropped after class t training. So having an ordered list of exemplars according to importance is a key to LL since it is easy to reduce its size with future new classes added while retaining the most essential past information.

Algorithm 4.4 iCaRL ConstructExemplarSet

Input: examples $X = \{x_1, \ldots, x_n\}$ of class t, the target number of exemplars m, current feature function $\varphi : \mathcal{X} \to \mathbb{R}^d$.
Output: exemplar set P for class y.

1: $\mu \leftarrow \frac{1}{n} \sum_{x \in X} \varphi(x)$
2: **for** $k = 1$ **to** m **do**
3:     $p_k \leftarrow \text{argmin}_{x \in X \text{ and } x \notin \{p_1, \ldots, p_{k-1}\}} \left\| \mu - \frac{1}{k} [\varphi(x) + \sum_{j=1}^{k-1} \varphi(p_j)] \right\|$
4: **end for**
5: $P \leftarrow (p_1, \ldots, p_m)$

4.6.4 PERFORMING CLASSIFICATION IN ICARL

With all the training algorithms introduced above, the classification is performed with the sets of exemplars $\mathcal{P} = \{P_1, \ldots, P_t\}$. The idea is straightforward: given a test example x, we pick the class y^* whose exemplar set's average feature vector is the closet to x as x's class label (see Algorithm 4.5).

Algorithm 4.5 iCaRL Classify in iCaRL

Input: a test example x to be classified, sets of exemplars $\mathcal{P} = \{P_1, \ldots, P_t\}$, current feature function $\varphi : \mathcal{X} \to \mathbb{R}^d$.
Output: predicted class label y^* of x.

1: **for** $y = 1$ **to** t **do**
2:     $\mu_y \leftarrow \frac{1}{|P_y|} \sum_{p \in P_y} \varphi(p)$
3: **end for**
4: $y^* \leftarrow \underset{y=1,\ldots,t}{\text{argmin}} \|\varphi(x) - \mu_y\|$

4.7   EXPERT GATE

Aljundi et al. [2016] proposed a *Network of Experts* where each expert is a model trained given a specific task. Since an expert is trained on one task only, it is good at this particular task, but not others. Thus, in the LL context, a network of experts are needed to handle a sequence of tasks.

One compelling point that Aljundi et al. [2016] emphasizes is the importance of *memory efficiency*, especially in the era of big data. As we know, GPUs are widely used for training deep learning models due to their rapid processing capability. However, GPUs have limited memory compared to CPUs. As deep learning models are becoming more and more complex, GPUs can only load a small number of models at a time. With a large number of tasks, as in LL, it requires the system to know what model or models to load when making a prediction on a test example.

With this need in mind, Aljundi et al. [2016] proposed an *Expert Gate* algorithm to determine the relevance of tasks, and only load the most relevant tasks in memory during inference. Denoting the existing tasks as $\mathcal{T}_1, \mathcal{T}_2, \ldots, \mathcal{T}_N$, an undercomplete autoencoder model A_k and an expert model E_k are constructed for each existing task $\mathcal{T}_k$ where $k \in \{1, \ldots, N\}$. When a new task $\mathcal{T}_{N+1}$ and its training data $\mathcal{D}_{N+1}$ arrive, $\mathcal{D}_{N+1}$ will be evaluated against each autoencoder A_k to find the most relevant tasks. The expert models of these most relevant tasks are used for fine-tuning or learning-without-forgetting (LwF) (Section 4.3) to build the expert model E_{N+1}. At the same time, A_{N+1} is learned from $\mathcal{D}_{N+1}$. When making a prediction on a test example x_t, the expert models whose corresponding autoencoders best describe x_t are loaded in memory and used to make the prediction.

4.7.1 AUTOENCODER GATE

An autoencoder [Bourlard and Kamp, 1988] model is a neural network that learns to recover input in the output layer in an unsupervised manner. There are encoder and decoder in the model. The encoder $f = h(x)$ projects the input x to an embedded space $h(x)$ while the decoder $r = g(h(x))$ maps the embedded space to the original input space. There are two types of autoencoder models: undercomplete autoencoder and overcomplete autoencoder. Undercomplete autoencoder learns a lower-dimensional representation and overcomplete autoencoder learns a higher-dimensional representation with regularization. An example of undercomplete autoencoder is illustrated in Figure 4.2.

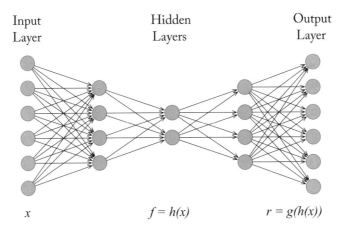

Figure 4.2: An example of undercomplete autoencoder model.

The intuition for using autoencoder in Expert Gate is that, as an unsupervised approach, an undercomplete autoencoder can learn a lower-dimensional feature representation that best describes the data in a compact way. The autoencoder of one task should perform well at reconstructing the data of that task, i.e., one autoencoder model is a decent representation of one task. If two autoencoder models from two tasks are close to each other, the tasks are likely to be similar too.

The autoencoder used in Aljundi et al. [2016] is simple: it has one ReLU layer [Zeiler et al., 2013] between the encoding and decoding layers. ReLU activation units are fast and easy to optimize, which also introduce sparsity to avoid over-fitting.

4.7.2   MEASURING TASK RELATEDNESS FOR TRAINING

Given a new task $\mathcal{T}_{N+1}$ with its training data $\mathcal{D}_{N+1}$, Expert Gate first learns an autoencoder A_{N+1} from $\mathcal{D}_{N+1}$. To facilitate training expert model E_{N+1}, it finds the most related existing task and use its expert model. Specifically, the reconstruction error of $\mathcal{D}$ using an autoencoder A_k is defined as:

$$Er_k = \frac{\sum_{x \in \mathcal{D}} er_x^k}{|\mathcal{D}|} \; , \qquad (4.8)$$

where er_x^k is the reconstruction error of applying x to the autoencoder A_k. Since the data of existing tasks are discarded, only $\mathcal{D}_{N+1}$ can be used to evaluate the relatedness. Given an existing task $\mathcal{T}_k$, $\mathcal{D}_{N+1}$ is used to compute two reconstruction errors: Er_{N+1} of autoencoder A_{N+1} and E_k of autoencoder A_k. The task relatedness is thus defined as:

$$Relatedness(\mathcal{T}_{N+1}, \mathcal{T}_k) = 1 - \frac{Er_{N+1} - Er_k}{Er_k} \; . \qquad (4.9)$$

Note that this relatedness definition is asymmetric. After the most related task is chosen, depending on how related it is to the new task, fine-tuning (see Section 4.1) or learning-without-forgetting (LwF) (Section 4.3) is employed. If two tasks are sufficiently related, LwF is applied; otherwise, fine-tuning is used. In LwF, a shared model is used for all tasks while each task has its own classification layer. A new task introduces a new classification layer. The model is fine-tuned to the new task's data while trying to preserve the previous tasks' predictions on the new data.

4.7.3   SELECTING THE MOST RELEVANT EXPERT FOR TESTING

If a test example x_t yields a very small reconstruction error when going through an autoencoder (say A_k), x_t should be similar to the data that was used to train A_k. The specialized model (expert) E_k should hence be utilized to make predictions on x_t. The probability p_k of x_t being relevant to an expert E_k is defined as:

$$p_k = \frac{exp(-er_{x_t}^k / \tau)}{\sum_j exp(-er_{x_t}^j / \tau)} \; , \qquad (4.10)$$

where $er_{x_t}^k$ is the reconstruction error of applying x_t to the autoencoder A_k. τ is the temperature whose value is 2, leading to soft probability values. Aljundi et al. [2016] picked the expert $E_{k'}$

to make the prediction on x_t whose $p_{k'}$ is the maximum among all existing tasks. The approach can also accommodate loading multiple experts by simply selecting experts whose relevant score is higher than a threshold.

4.7.4   ENCODER-BASED LIFELONG LEARNING

Finally, we note that Rannen Ep Triki et al. [2017] also used the idea of autoencoder to extend LwF (Section 4.3). Rannen Ep Triki et al. [2017] argued that LwF has an inferior loss function definition when the new task data distribution is quite different from those of previous tasks. To address this issue, an autoencoder based method is proposed to preserve only the features that are the most important for previous tasks while allowing other features to adapt more quickly to new tasks. This is achieved by learning a lower-dimensional manifold via autoencoder, and constraining the distance between the reconstructions. Note that this is similar to EWC (Section 4.5) in the sense that EWC tries to maintain the most important weights while Rannen Ep Triki et al. [2017] aims to conserve features. See Rannen Ep Triki et al. [2017] for more details.

4.8   CONTINUAL LEARNING WITH GENERATIVE REPLAY

Shin et al. [2017] proposed a continual learning method using replayed examples from a generative model without referring to the actual data of past tasks. It is inspired by the suggestion that the hippocampus is better paralleled with a generative model than a replay buffer [Ramirez et al., 2013, Stickgold and Walker, 2007]. As mentioned in Section 4.2, this represents a stream of lifelong learning systems that use dual-memory for knowledge consolidation. We pick the work of Shin et al. [2017] to give a flavor of such models. In the deep generative replay framework proposed by Shin et al. [2017], a generative model is maintained to feed pseudo-data as knowledge of past tasks to the system. To train such a generative model, the generative adversarial networks (GANs) [Goodfellow et al., 2014] framework is used. Given a sequence of tasks, a *scholar* model, containing a generator and a solver, is learned and retained. Such a scholar model holds the knowledge representing the previous tasks, and thus prevents the system from forgetting previous tasks.

4.8.1   GENERATIVE ADVERSARIAL NETWORKS

The Generative Adversarial Networks (GANs) framework is not only used in Shin et al. [2017], but also widely adopted in the deep learning community (e.g., Radford et al. [2015]). In this subsection, we give an overview of GANs based on Goodfellow [2016].

In GANs, there are two players: a *generator* and a *discriminator*. On the one hand, the generator creates samples that mimic training data, i.e., drawing samples from the similar (ideally same) distribution as the training data. On the other hand, the discriminator classifies the samples to tell whether they are real (from real training data) or fake (from samples created by the generator). The problem that discriminator faces is a typical binary classification problem. Following the example given in Goodfellow [2016], a generator is like a counterfeiter who tries to make fake money. A discriminator is like a police who wants to allow legitimate money and catch counterfeit money. To win the game, the counterfeiter (generator) must learn how to make money that looks identical to genuine money while the police (discriminator) learns how to distinguish authenticity without mistakes.

Formally, GANs are a structured probabilistic model with latent variables z and observed variables x. The discriminator has a function D that takes x as input. The function for the generator is defined as G whose input is z. Both functions are differentiable with respect to their inputs and parameters. The cost function for the discriminator is:

$$J = -\frac{1}{2}\mathbb{E}_{x \sim p_{data}(x)}[\log D(x)] - \frac{1}{2}\mathbb{E}_{z \sim p_z(z)}[\log(1 - D(G(z)))] \ . \tag{4.11}$$

By treating the two-player game as a *zero-sum game* (or *minimax* game), the solution involves minimization in an outer loop and maximization in an inner loop, yielding the objective function for discriminator D and generator G as:

$$\begin{aligned} \mathcal{L}(D, G) &= \min_G \max_D V(D, G) \\ &= \min_G \max_D -J \\ &= \min_G \max_D \mathbb{E}_{x \sim p_{data}(x)}[\log D(x)] + \mathbb{E}_{z \sim p_z(z)}[\log(1 - D(G(z)))] \ . \end{aligned} \tag{4.12}$$

4.8.2   GENERATIVE REPLAY

In Shin et al. [2017], a *scholar* model H is learned and maintained in an LL manner. The scholar model contains a generator G and a solver S with parameters θ. The solver here is like the discriminator in Section 4.8.1. Denoting the previous N tasks as $\mathcal{T}_N = (\mathcal{T}_1, \mathcal{T}_2, \dots, \mathcal{T}_N)$, and the scholar model for previous N task as $H_N = \langle G_N, S_N \rangle$, the system aims to learn a new scholar model $H_{N+1} = \langle G_{N+1}, S_{N+1} \rangle$ given the new task $\mathcal{T}_{N+1}$'s training data $\mathcal{D}_{N+1}$.

To obtain $H_{N+1} = \langle G_{N+1}, S_{N+1} \rangle$ given the training data $\mathcal{D}_{N+1} = (x, y)$, there are two steps.

1. G_{N+1} is updated with the new task input x and replayed inputs x' created from G_N. Real and replayed samples are mixed at a ratio that depends on the importance of the new task compared to previous ones. Recall that this step is known as intrinsic replay or pseudo-rehearsal [Robins, 1995] in which new data and replayed samples of old data are mixed to prevent catastrophic forgetting.

2. S_{N+1} is trained to couple the inputs and targets drawn from the same mix of real and replayed data, with the loss function:

$$\begin{aligned} \mathcal{L}_{train}(\theta_{N+1}) = \ & r\mathbb{E}_{(x,y) \sim \mathcal{D}_{N+1}}[L(S(x; \theta_{N+1}), y)] \\ & + (1 - r)\mathbb{E}_{x' \sim G_N}[L(S(x'; \theta_{N+1}), S(x'; \theta_N))] \ , \end{aligned} \tag{4.13}$$

where θ_N denotes the parameters for the solver S_N, and r denotes the ratio of mixing real data. If S_N is tested on the previous tasks, the test loss function becomes:

$$\mathcal{L}_{test}(\theta_{N+1}) = r\mathbb{E}_{(\boldsymbol{x},\boldsymbol{y})\sim\mathcal{D}_{N+1}}[L(S(\boldsymbol{x};\theta_{N+1}),\boldsymbol{y})]$$
$$+ (1-r)\mathbb{E}_{(\boldsymbol{x},\boldsymbol{y})\sim\mathcal{D}_{past}}[L(S(\boldsymbol{x};\theta_{N+1}),\boldsymbol{y})] \ , \qquad (4.14)$$

where $\mathcal{D}_{past}$ is the cumulative distribution of the data from the past tasks.

The proposed framework is independent of any specific generative model or solver. The choice for the deep generative model can be a variational autoencoder [Kingma and Welling, 2013] or a GAN [Goodfellow et al., 2014].

4.9 EVALUATING CATASTROPHIC FORGETTING

There are two main papers [Goodfellow et al., 2013a, Kemker et al., 2018] in the literature that evaluate ideas aimed at addressing catastrophic forgetting in neural networks.

Goodfellow et al. [2013a] evaluated some traditional approaches that attempt to reduce catastrophic forgetting. They evaluated dropout training [Hinton et al., 2012] as well as various activation functions including:

- logistic sigmoid,

- rectified linear [Jarrett et al., 2009],

- hard local winner take all (LWTA) [Srivastava et al., 2013], and

- Maxout [Goodfellow et al., 2013b].

They also used random hyperparameter search [Bergstra and Bengio, 2012] to automatically select hyperparameters. In terms of experiments, only pairs of tasks were considered in Goodfellow et al. [2013a] with one being the "old task" and the other being the "new task." The tasks were MNIST classification [LeCun et al., 1998] and sentiment classification on Amazon reviews [Blitzer et al., 2007]. Their experiments showed that dropout training is mostly beneficial to prevent forgetting. They also found that the choice of activation function matters less than the choice of training algorithm.

Kemker et al. [2018] evaluated several more recent continual learning algorithms using larger datasets. These algorithms include the following.

- Elastic weight consolidation (EWC) [Kirkpatrick et al., 2017]: it reduces plasticity of important weights with respect to previous tasks when adapting to a new task (see Section 4.5).

- PathNet [Fernando et al., 2017]: it creates an independent output layer for each task to preserve previous tasks. It also finds the optimal path to be trained when learning a particular task, which is like a dropout network.

- GeppNet [Gepperth and Karaoguz, 2016]: it reserves a sample set of training data of previous tasks, which is replayed to serve as a short-term memory when training on a new task.

- Fixed expansion layer (FEL) [Coop et al., 2013]: it uses sparsity in representation to mitigate catastrophic forgetting.

They proposed three benchmark experiments for measuring catastrophic forgetting.

1. **Data Permutation Experiment**: The elements in the feature vector are randomly permutated. In the same task, the permutation order is the same while different tasks have distinct permutation orders. This is similar to the experiment setup in Kirkpatrick et al. [2017].

2. **Incremental Class Learning**: After learning the base task set, each new task contains only a single class to be incrementally learned.

3. **Multi-Modal Learning**: The tasks contain different datasets, e.g., learn image classification and then audio classification.

Three datasets were used in the experiments: MNIST [LeCun et al., 1998], CUB-200 [Welinder et al., 2010], and AudioSet [Gemmeke et al., 2017]. Kemker et al. [2018] evaluated the accuracy on the new task as well as the old tasks in the LL setting, i.e., tasks arriving in a sequence. They found that PathNet performs the best in data permutation, GreppNet obtains the best accuracy in incremental class learning, and EWC has the best results in multi-modal learning.

4.10   SUMMARY AND EVALUATION DATASETS

This chapter reviewed the problem of catastrophic forgetting and existing continual learning algorithms aimed at dealing with it. Most existing works fall into some variations of regularization or increasing/allocating extra parameters for new tasks. They are shown to be effective in some simplified LL settings. Considering the huge success of deep learning in recent years, continual/lifelong deep learning continues to be one of the most promising channels to reach true intelligence with embedded LL. Nonetheless, catastrophic forgetting remains a long-standing challenge. We look forward to the day when a robot can learn to perform all kinds of tasks and solve all kinds of problems continually and seamlessly without human intervention and without interfering each other.

To reach this ideal, there are many obstacles and gaps. We believe that one major gap is how to seamlessly discover, integrate, organize, and solve problems or tasks of different similarities at the different levels of detail in a single network or even multiple networks just like our human brains, with minimum interference of each other. For example, some tasks are dissimilar at the detailed action level but may be similar at a higher or more abstract level. How to automatically recognize and leverage the similarities and differences in order to learn quickly and better in an incremental and lifelong manner without the need of a large amount of training data is a very challenging and interesting research problem.

Another gap is the lack of research in designing systems that can truly embrace real-life problems with memories. This is particularly relevant to DNNs due to catastrophic forgetting. One idea is to encourage the system to take snapshots of its status and parameters, and keep validating itself against a gold dataset. It is not practical to retain all the training data. But to prevent the system from moving to some extreme parameter point in the space, it is useful to keep a small sampled set of training data that can cover most of the patterns/classes seen before.

In short, catastrophic forgetting is a key challenge for DNNs to enable LL. We hope this chapter can shed some light in the area and attract more attention to address this challenge.

Regarding evaluation datasets, image data are among the most commonly used datasets for evaluating continual learning due to their wide availability. Some of the common ones are as follows.

- **MNIST** [LeCun et al., 1998][1] is perhaps the most commonly used dataset (used in more than half of the works introduced in this chapter). It consists of labeled examples of hand-written digits. There are 10 digit classes. One way to produce datasets for multiple tasks is to create the representations of the data by randomly permuting the elements of input feature vectors [Goodfellow et al., 2013a, Kemker et al., 2018, Kirkpatrick et al., 2017]. This paradigm ensures that the tasks are overlapping and have equal complexity.

- **CUB-200** (Caltech-UCSD Birds 200) [Welinder et al., 2010][2]) is another popular dataset for LL evaluation. It is an image dataset with photos of 200 bird species. It has been used in Aljundi et al. [2016, 2017], Kemker et al. [2018], Li and Hoiem [2016], Rannen Ep Triki et al. [2017], and Rosenfeld and Tsotsos [2017].

- **CIFAR-10** and **CIFAR-100** [Krizhevsky and Hinton, 2009][3] are also widely used. They contain images of 10 classes and 100 classes, respectively. They are used in Fernando et al. [2017], Jung et al. [2016], Lopez-Paz et al. [2017], Rebuffi et al. [2017], Venkatesan et al. [2017], Zenke et al. [2017], and Rosenfeld and Tsotsos [2017].

- **SVHN** (Google Street View House Numbers) [Netzer et al., 2011][4] is similar to MNIST, but contains an order of magnitude more labeled data. These images are from real-world problems and are harder to solve. It also has 10 digit classes. It is used in Aljundi et al. [2016, 2017], Fernando et al. [2017], Jung et al. [2016], Rosenfeld and Tsotsos [2017], Shin et al. [2017], Venkatesan et al. [2017], and Seff et al. [2017].

[1]http://yann.lecun.com/exdb/mnist/
[2]http://www.vision.caltech.edu/visipedia/CUB-200.html
[3]https://www.cs.toronto.edu/~kriz/cifar.html
[4]http://ufldl.stanford.edu/housenumbers/

Other image datasets include **Caltech-256** [Griffin et al., 2007],[5] **GTSR** [Stallkamp et al., 2012],[6] **Human Sketch dataset** [Eitz et al., 2012],[7] **Daimler** (DPed) [Munder and Gavrila, 2006],[8] **MIT Scenes** [Quattoni and Torralba, 2009],[9] **Flower** [Nilsback and Zisserman, 2008],[10] **FGVC-Aircraft** [Maji et al., 2013],[11] **ImageNet ILSVRC2012** [Russakovsky et al., 2015],[12] and **Letters** (Chars74K) [de Campos et al., 2009].[13]

More recently, Lomonaco and Maltoni [2017] proposed a dataset called **CORe50**.[14] It contains 50 objects that were collected in 11 distinct sessions (8 indoor and 3 outdoor) differing in background and lighting. The dataset is specifically designed for continual object recognition. Unlike many popular datasets such as **MNIST** and **SVHN**, **CORe50**'s multiple views of the same object from different sessions enable richer and more practical LL. Using **CORe50**, Lomonaco and Maltoni [2017] considered evaluation settings where the new data can contain (1) new patterns of the existing classes, (2) new classes, and (3) new patterns and new classes. Such real-life evaluation scenarios are very useful for carrying the LL research forward. Parisi et al. [2018b] used **CORe50** to perform an evaluation of their own approach as well as some other approaches, e.g., LwF [Li and Hoiem, 2016], EWC [Kirkpatrick et al., 2017], and iCaRL [Rebuffi et al., 2017].

Apart from image datasets, some other types of data are also used. **AudioSet** [Gemmeke et al., 2017][15] is a large-scale collection of human-labeled 10-sec sound clips sampled from YouTube videos. It is used in Kemker et al. [2018].

In continual learning on reinforcement learning, different environments were used for evaluation. **Atari games** [Mnih et al., 2013] are among the most popular ones which are used in Kirkpatrick et al. [2017], Rusu et al. [2016], and Lipton et al. [2016]. Some other environments include **Adventure Seeker** [Lipton et al., 2016], **CartPole-v0** in OpenAI gym [Brockman et al., 2016], and **Treasure World** [Mankowitz et al., 2018].

---

[5] http://ufldl.stanford.edu/housenumbers/
[6] http://benchmark.ini.rub.de/
[7] http://cybertron.cg.tu-berlin.de/eitz/projects/classifysketch/
[8] http://www.gavrila.net/Datasets/Daimler_Pedestrian_Benchmark_D/daimler_pedestrian_benchmark_d.html
[9] http://web.mit.edu/torralba/www/indoor.html
[10] http://www.robots.ox.ac.uk/~vgg/data/flowers/
[11] http://www.robots.ox.ac.uk/~vgg/data/fgvc-aircraft/
[12] http://www.image-net.org/challenges/LSVRC/
[13] http://www.ee.surrey.ac.uk/CVSSP/demos/chars74k/
[14] https://vlomonaco.github.io/core50/benchmarks.html
[15] https://research.google.com/audioset/dataset/index.html

CHAPTER 5

Open-World Learning

Classic supervised learning makes the *closed-world assumption*, meaning that all the test classes have been seen in training [Bendale and Boult, 2015, Fei and Liu, 2016, Fei et al., 2016]. Although this assumption holds in many applications, it is violated in many others, especially in dynamic and open environments, where instances of unexpected classes may appear in testing or applications. That is, the test/application data may contain instances from classes that have not appeared in training. To learn in such an environment, we need *open-world learning* (open-world classification or simply open classification), which has to detect instances of unseen classes during testing or model application, and incrementally learn the new classes to update the existing model without re-training the whole model from scratch. This form of learning is also called *cumulative learning* in Fei et al. [2016] and in the first edition of this book. In computer vision, open-world learning is called *open-world recognition* [Bendale and Boult, 2015, De Rosa et al., 2016].

In fact, open-world learning is a general problem, not limited to supervised learning. It can be broadly defined as learning a model that can perform its intended task and also identify new things that have not been learned before, and then incrementally learn the new things. Open-world learning can occur in different learning scenarios and paradigms. For example, in reading, the system may see a new word that it does not know, and then learns it by looking up the word in the dictionary. In human-machine conversation, the conversation agent may not understand something said by the human user and then asks the user to explain in order to learn it. In this chapter, we focus on open-world supervised learning. Learning during conversation will be discussed in Chapter 8.

Open-world learning basically performs a form of *self-motivated learning* because by recognizing that something new has appeared, the system has the opportunity to learn the new thing. Traditionally, self-motivated learning means that the learner has curiosity that motivates it to explore new territories and to learn new things. In the context of supervised learning, the key is for the system to recognize what it has not seen or learned before. If a learned model cannot recognize any new things, there is no way for the learner to learn new things or to explore by itself other than by being told or instructed by a human user or an external system, which is not ideal for a truly intelligent system. It also has great difficulty to function in a dynamic and open environment.

5.1   PROBLEM DEFINITION AND APPLICATIONS

Open-world Learning is defined as follows [Bendale and Boult, 2015, Fei et al., 2016].

1. At a particular time point, the learner has built a multi-class classification model F_N based on all past N classes of data $\mathcal{D}^p = \{\mathcal{D}_1, \mathcal{D}_2, \ldots, \mathcal{D}_N\}$ with their corresponding class labels $\mathcal{Y}^N = \{l_1, l_2, \ldots, l_N\}$. F_N is able to classify each test instance to either one of the known classes $l_i \in \mathcal{Y}^N$ or reject it and put it in a rejected set R, which may include instances from one or more new or unseen classes in the test set.

2. The system or a human user identifies the hidden unseen classes C in R, and collects training data for the unseen classes.

3. Assume that there are k new classes in C that have enough training data. The learner incrementally learns the k classes based on their training data. The existing model F_N is updated to produce the new model F_{N+k}.

Open-world learning is a form of lifelong learning (LL) because it conforms to the definition of LL in Chapter 1. Specifically, the new learning task $\mathcal{T}_{N+1}$ is to build a multi-class open classifier based on all the past and the current classes. The knowledge base (KB) contains the past model F_N and possibly all the past training data.

   We should note that the third task of learning new classes incrementally here is different from traditional *incremental class learning* (ICL) studied in different areas because traditional ICL still learns in the closed-world (i.e., it does not perform unseen class rejection) although it can add new classes incrementally to the classification system without re-training the whole model from scratch.

   Let us see some example applications. For example, we want to build a greeting robot for a hotel. At any point in time, the robot has learned to recognize all existing hotel guests. When it sees an existing guest, it can call his/her name and chat. At the same time, it must also detect any new guests that it has not seen before. On seeing a new guest, it can say hello, ask for his/her name, take many pictures, and learn to recognize the guest. Next time when it sees the person again, it can call his/her name and chat like an old friend. The scenario in self-driving cars is very similar as it is very hard, if not impossible, to train a system to recognize every possible object that may appear on the road. The system has to recognize objects that it has not learned before and learn them during driving (possibly through interactions with the human passenger) so that when it sees the objects next time, it will have no problem recognizing them.

   Fei et al. [2016] gave another example in text classification. The 2016 presidential election in the U.S. was a hot topic on social media, and many social science researchers relied on collected online user discussions to carry out their research. During the campaign, every new proposal made by a candidate was followed by a huge amount of discussions in the social media. A multi-class classifier is thus needed to organize the discussions. As the campaign went on, the initially built classifier inevitably encounters new topics (e.g., Donald Trump's plan for immigration reform or Hillary Clinton's proposal for tax increase) that had not been covered in previous training. In this case, the classifier should first recognize these new topics when they occur rather than classify them into some existing classes or topics. Second, after enough training examples of the new topics are collected, the existing classifier should incorporate the new classes or topics incrementally in a manner that does not require retraining the entire classification system from scratch.

Bendale and Boult [2015] made an attempt to solve the open-world learning problem (which was called open-world recognition in their paper) for image classification. Its method is called *Nearest Non-Outlier* (NNO), which is adapted from the traditional *Nearest Class Mean* (NCM) method for image classification using a metric learning technique proposed by Mensink et al. [2013]. In NCM, each image is represented as a feature vector and each class is represented by the class mean computed using the feature vectors of all the images in the class. In testing, each test image's feature vector is compared with each class mean and is assigned the class with the nearest class mean. However, this method cannot perform unseen class rejection. NNO enables rejection. For incremental learning, it simply adds the new class mean to the existing class mean set. The rejection capability of NNO was improved in Bendale and Boult [2016]. The new method, called OpenMax, is based on deep learning, which adapts the traditional SoftMax classification scheme to enable rejection by introducing a new model layer (also called OpenMax) to estimate the probability of an input being from an unseen class. However, its training needs examples from some unseen classes (not necessarily the test unseen classes) to tune the parameters. In the next two sections, we discuss two other methods. It was shown in Shu et al. [2017a] that its DOC method outperforms OpenMax for both open text and open image classifications without requiring any training unseen class examples.

5.2   CENTER-BASED SIMILARITY SPACE LEARNING

Fei et al. [2016] proposed a technique to perform open-world classification based on a center-based similarity space learning method (called *CBS learning*), which we discuss below. We first discuss its training process for learning a new class incrementally and then its testing process, which is able to classify test instances to known/seen classes and also detect unseen class instances.

5.2.1   INCREMENTALLY UPDATING A CBS LEARNING MODEL

This sub-section describes incremental training in CBS learning, which was inspired by human concept learning. Humans are exposed to new concepts all the time. One way we learn a new concept is perhaps by searching from the already known concepts for ones that are similar to the new concept, and then trying to find the difference between these known concepts and the new one without using all the known concepts. For example, assume we have already learned the concepts like "movie," "furniture," and "soccer." Now we are presented with the concept of "basketball" and its set of documents. We find that "basketball" is similar to "soccer," but very different from "movie" and "furniture." Then we just need to *accommodate* the new concept "basketball" into our old knowledge base by focusing on distinguishing the "basketball" and "soccer" concepts, and do not need to worry about the difference between "basketball" and "movie" or "furniture," because the concepts of "movie" and "furniture" can easily tell that documents from "basketball" do not belong to either of them.

Fei et al. [2016] adopted this idea and used the 1-vs.-rest strategy of SVM for incremental learning of multiple classes (or concepts). Before the new class l_{N+1} arrives, the learning system has built a classification model F_N, which consists of a set of N 1-vs.-rest binary classifiers $F_N = \{f_1, f_2, \ldots, f_N\}$, for the past N classes using their training sets $\mathcal{D}^p = \{\mathcal{D}_1, \mathcal{D}_2, \ldots, \mathcal{D}_N\}$ and corresponding class labels $\mathcal{Y}^N = \{l_1, l_2, \ldots, l_N\}$. Each f_i is a binary classifier built using the CBS learning method (see Section 5.2.3) for identifying instances of class l_i. When a new dataset $\mathcal{D}_{N+1}$ of class l_{N+1} arrives, the system goes through the following two steps to update the classification model F_N to build a new model F_{N+1} in order to be able to classify test data or instances of all existing classes in $\mathcal{Y}^{N+1} = \{l_1, l_2, \ldots, l_N, l_{N+1}\}$ and recognize any unseen class C_0 of documents.

1. Searching for a set of classes SC that are similar to the new class l_{N+1}.

2. Learning to separate the new class l_{N+1} and the previous classes in SC.

For step 1, the similarity between the new class l_{N+1} and the previous ones $\{l_1, l_2, \ldots, l_N\}$ is computed by running each of the 1-vs.-rest past binary classifiers f_i in $F_N = \{f_1, f_2, \ldots, f_N\}$ to classify instances in $\mathcal{D}_{N+1}$. The classes of those past binary classifiers that accept (classify as positive) a certain number/percentage λ_{sim} of instances from $\mathcal{D}_{N+1}$ are regarded as similar classes and denoted by SC.

Step 2 of separating the new class l_{N+1} and classes in SC involves two sub-steps: (1) building a new binary classifier f_{N+1} for the new class l_{N+1} and (2) updating the existing classifiers for the classes in SC. It is intuitive to build f_{N+1} using $\mathcal{D}_{N+1}$ as the positive training data and the data of the classes in SC as the negative training data. The reason for updating classifiers in SC is that the joining of class l_{N+1} confuses those classifiers in SC. To re-build each classifier, the system needs to use the original negative data employed to build the existing classifier f_i and the new data $\mathcal{D}_{N+1}$ as the new negative training data. The reason that the old negative training data is still used is because the new classifier still needs to separate class l_i from those old classes.

The detailed algorithm is given in Algorithm 5.6, which incrementally learns a new class with its data $\mathcal{D}_{N+1}$. Line 1 initializes SC to the empty set. Line 3 initializes the variable CT (count) to record the number of instances in $\mathcal{D}_{N+1}$ that will be classified as positive by classifier f_i. Lines 4–9 use f_i to classify each instance in $\mathcal{D}_{N+1}$ and record the number of instances that are classified (or accepted) as positive by f_i. Lines 10–12 check whether there are too many instances in $\mathcal{D}_{N+1}$ that have been classified as positive by f_i to render class l_i as similar to class l_{N+1}. λ_{sim} is a threshold controlling how many percents of instances in $\mathcal{D}_{N+1}$ should be classified to class l_i before considering l_i as similar/close to class l_{N+1}. Lines 14–17 build a new classifier f_{N+1} and update all the classifiers for classes in SC.

Algorithm 5.6 Incremental Class Learning

Input: classification model $F_N = \{f_1, f_2, \ldots, f_N\}$, past datasets $\{\mathcal{D}_1, \mathcal{D}_2, \ldots, \mathcal{D}_N\}$, new dataset $\mathcal{D}_{N+1}$, similarity threshold λ_{sim}.
Output: classification model $F_{N+1} = \{f_1, \ldots, f_N, f_{N+1}\}$

1: $SC = \emptyset$
2: **for** each classifier $f_i \in F_N$ **do**
3:      $CT = 0$
4:      **for** each test instance $x_j \in \mathcal{D}_{N+1}$ **do**
5:          $class = f_i(x_j)$ // classify document x_j using f_i
6:          **if** $class = l_i$ **then**
7:              $CT = CT + 1$ // wrongly classified
8:          **end if**
9:      **end for**
10:     **if** $CT > \lambda_{sim} \times |\mathcal{D}_{N+1}|$ **then**
11:         $SC = SC \cup \{l_i\}$
12:     **end if**
13: **end for**
14: Build f_{N+1} and add it to F_{N+1}
15: **for** each f_i of class $l_i \in SC$ **do**
16:     Update f_i
17: **end for**
18: Return F_{N+1}

In summary, the learning process uses the set SC of similar classes to the new class l_{N+1} to control both the number of binary classifiers that need to be built/updated and also the number of negative instances used in building the new classifier f_{N+1}. It thus greatly improves the efficiency compared to building a new multi-class classifier F_{N+1} from scratch.

Combining the above incremental learning process and the underlying classifier *cbsSVM* discussed in Section 5.2.3, the new learner, called *CL-cbsSVM* (*CL* stands for *Cumulative Learning*, a name used in Fei et al. [2016] for open-world learning) is able to tackle both challenges in incremental learning.

5.2.2   TESTING A CBS LEARNING MODEL

To test the new classification model $F_{N+1} = \{f_1, f_2, \ldots, f_N, f_{N+1}\}$, the standard technique of combining the set of 1-vs.-rest binary classifiers to perform multi-class classification is followed with a rejection option for the unknown. As output scores from different SVM classifiers are not comparable, the SVM scores for each classifier are first converted to probabilities based on a variant of Platt's algorithm [Platt et al., 1999], which is supported in LIBSVM [Chang and Lin, 2011]. Let $P(y|x)$ be a probabilistic estimator, where $y \in Y^{N+1}$ $(= \{l_1, l_2, \ldots, l_N, l_{N+1}\})$ is a class label and x is the feature vector of a test instance. Let θ $(= 0.5)$ be the decision threshold, y^* be the final predicted class for x, and C_0 be the label for the unknown. Classification of the test instance x is done as follows:

$$y^* = \begin{cases} \mathrm{argmax}_{y \in Y^{N+1}} \, P(y|x) & \text{if } P(y|x) \geq \theta \\ C_0 & \text{otherwise} \end{cases}. \tag{5.1}$$

The idea is that for the test instance x, each binary classifier $f_i \in F_{N+1}$ is used to produce a probability $P(l_i|x)$. If none of the probabilities is greater than θ $(= 0.5)$, the document represented by x is regarded as unseen/unknown and belonging to C_0; otherwise it is classified to the class with the highest probability.

5.2.3   CBS LEARNING FOR UNSEEN CLASS DETECTION

This subsection describes the core CBS learning method, which performs binary classification focusing on identifying positive class documents and also has the ability to detect unseen classes or classifying them as not positive. It provides the base learning method for open-world learning above [Fei et al., 2016]. The learning method is based on the idea of reducing the *open space risk* while balancing the *empirical risk* in learning. Classic learners define and optimize over empirical risk, which is measured on the training data. For open learning, it is crucial to consider how to extend the classic model to capture the risk of the unknown by preventing over-generalization. To tackle this problem, Scheirer et al. [2013] introduced the concept of *open space risk*. Below, we first discuss the open space risk management strategy in Fei et al. [2016], and then apply an SVM-based CBS learning method as the solution toward managing the open space risk. The basic idea of CBS learning is to find a "ball" (decision boundary) to cover the positive class data area. Any document falling outside of the "ball" is considered not positive. Although CBS learning only performs binary classification, applying the 1-vs.-rest method described in Section 5.2.2 gives a multi-class CBS classification model, which is called cbsSVM in Fei et al. [2016].

Open Space Risk
Consider the risk formulation for open image recognition in Scheirer et al. [2013], where apart from empirical risk, there is risk in labeling the open space (space away from positive training examples) as "positive" for any unknown class. Due to lack of information of a classification function on the open space, open space risk is approximated by a relative Lebesgue measure [Shackel, 2007]. Let S_o be a large ball of radius r_o that contains both the positively labeled open space O and all of the positive training examples; and let f be a measurable classification

function, where $f_y(x) = 1$ means recognizing x as belonging to class y of interest and $f_y(x) = 0$ otherwise. In our case, y is simply any class of interest l_i.

In Fei et al. [2016], O is defined as the positively labeled area that is sufficiently far from the center of the positive training examples. Let $B_{r_y}(cen_y)$ be a closed ball of radius r_y centered around the center cen_y of positive class y, which, ideally, should tightly cover all positive examples of class y only; S_o be a larger ball $B_{r_o}(cen_y)$ of radius r_o with the same center cen_y. Let classification function $f_y(x) = 1$ for $x \in B_{r_o}(cen_y)$, and $f_y(x) = 0$ otherwise. Also let q be the positive half space defined by a binary SVM decision hyperplane Ω obtained using positive and negative training examples. We also define the size of ball B_{r_o} to be bounded by Ω, $B_{r_o} \cap q = B_{r_o}$. Then the positive open space is defined as $O = S_o - Br_y(cen_y)$. S_o needs to be determined during learning for the positive class.

This open-space formulation greatly reduces the open space risk compared to traditional SVM and 1-vs.-Set Machine in Scheirer et al. [2013]. For traditional SVM, classification function $f_y^{SVM}(x) = 1$ when $x \in q$, and its positive open space is approximately $q - B_{r_y}(cen_y)$, which is only bounded by the SVM decision hyperplane Ω. For 1-vs.-Set Machine in Scheirer et al. [2013], $f_y^{1-vs-set}(x) = 1$ when $x \in g$, where g is a slab area with thickness δ bounded by two parallel hyperplanes Ω and Ψ ($\Psi \| \Omega$) in q. And its positive open space is approximately $g - g \cap B_{r_y}(cen_y)$. Given open-space formulations of the traditional SVM and 1-vs.-Set Machine, we can see that both methods label an unlimited area as the positively labeled space, while Fei et al. [2016] reduces it to a bounded area of a "ball."

Given the open space definition, the question is how to estimate S_o for the positive class. Fei et al. [2016] used the center-based similarity space learning (CBS learning), which transforms the original document space to a similarity space. The final classification is performed in the CBS space. Below, we introduce CBS learning and briefly discuss why it is suitable for the problem.

Center-Based Similarity Space Learning

Let $\mathcal{D} = \{(x_1, y_1), (x_2, y_2), \dots, (x_n, y_n)\}$ be the set of training examples, where x_k is the feature vector (e.g., with unigram features) representing a document and $y_k \in \{1, -1\}$ is its class label. This feature vector is called a document space vector (or *ds-vector*). Traditional classification directly uses $\mathcal{D}$ to build a binary classifier. However, CBS learning transforms each ds-vector x_k (no change to its class label) to a center-based similarity space feature vector (CBS vector) $cbs-v_k$. Each feature in $cbs-v_k$ is a similarity between a center c_j of the positive class documents and x_k.

To make CBS learning more effective by generating more similarity features, multiple document space representations or feature vectors (e.g., one based on unigrams and one based on bigrams) can be used to represent each document, which results in multiple centers for the positive documents. There can also be multiple document similarity functions used to compute similarity values. The detailed learning technique is as follows.

For a document x_k, we have a set R_k of p ds-vectors, $R_k = \{d_1^k, d_2^k, \dots, d_p^k\}$. Each ds-vector d_j^k denotes one document space representation of the document x_k, e.g., unigram representation or bigram representation. Then the centers of positive training documents can be computed, which are represented as a set of p centroids $C = \{c_1, c_2, \dots, c_p\}$. Each c_j corresponds to one document space representation in R_k. The Rocchio method in information retrieval [Manning et al., 2008] is used to compute each center c_j (a vector), which uses the corresponding ds-vectors of all training positive and negative documents:

$$c_j = \frac{\alpha}{|\mathcal{D}_+|} \sum_{x_k \in \mathcal{D}_+} \frac{d_j^k}{\left\| d_j^k \right\|} - \frac{\beta}{|\mathcal{D} - \mathcal{D}_+|} \sum_{x_k \in \mathcal{D} - \mathcal{D}_+} \frac{d_j^k}{\left\| d_j^k \right\|} , \qquad (5.2)$$

where $\mathcal{D}_+$ is the set of documents in the positive class and $|.|$ is the size function. α and β are parameters. It is reported that using the popular *tf-idf* (*term frequency and inverse document frequency*) representation, $\alpha = 16$ and $\beta = 4$ usually work well [Buckley et al., 1994]. The subtraction is used to reduce the influence of those terms that are not discriminative (i.e., terms appearing in both classes).

Based on R_k for a document x_k (in both training and testing) and the previously computed set C of centers using the training data, we can transform a document x_k from its document space representations R_k to one center-based similarity space vector $cbs\text{-}v_k$ by applying a similarity function *Sim* on each element d_j^k of R_k and its corresponding center c_j in C:

$$cbs\text{-}v_k = Sim(R_k, C) . \qquad (5.3)$$

Sim can contain a set of similarity measures. Each measure m is applied to p document representations d_j^k in R_k and their corresponding centers c_j in C to generate p similarity features (cbs-features) in $cbs\text{-}v_k$.

For ds-features, unigrams and bigrams with tf-idf weighting were used as two document representations. The five similarity measures in Fei and Liu [2015] were applied to measure the similarity of two vectors. Based on the CBS space representation, SVM is applied to produce a binary CBS classifier f_y.

Why Does CBS Learning Work?

We now briefly explain why CBS learning gives a good estimate to S_o. Due to using similarities as features, CBS learning generates a boundary to separate the positive and negative training data in the similarity space. Since similarity has no direction (or it covers all directions), the boundary in the similarity space is essentially a "ball" encompassing the positive class training data in the original document space. The "ball" is an estimate of S_o based on those similarity measures.

5.3   DOC: DEEP OPEN CLASSIFICATION

This section describes a deep learning based classification method called DOC [Shu et al., 2017a], which performs only classification and unseen class instance rejection, but does not do incremental learning of new classes. DOC is based on CNN [Collobert et al., 2011, Kim, 2014] and is augmented with a 1-vs.-rest final Sigmoid layer and Gaussian fitting for classification. This algorithm has been shown to perform better than many existing methods in both open-world text classification and open-world image classification.

5.3.1   FEED-FORWARD LAYERS AND THE 1-VS.-REST LAYER

The DOC system (given in Figure 5.1) is a variant of the CNN architecture [Collobert et al., 2011] for text classification [Kim, 2014].[1] The first layer embeds words in document x into dense vectors. The second layer performs convolution over dense vectors using different filters of varied sizes. Next, the max-over-time pooling layer selects the maximum values from the results of the convolution layer to form a k-dimension feature vector h. Then h is reduced to a N-dimension vector $d = d_{1:N}$ (N is the number of training/seen classes) via two fully connected layers and one intermediate ReLU activation layer:

$$d = W'(\text{ReLU}(Wh + b)) + b' , \tag{5.4}$$

where $W \in \mathbb{R}^{r \times k}$, $b \in \mathbb{R}^r$, $W' \in \mathbb{R}^{N \times r}$, and $b' \in \mathbb{R}^N$ are trainable weights; r is the output dimension of the first fully connected layer. The output layer of DOC is a 1-vs.-rest layer applied to $d_{1:N}$, which allows rejection. We describe it next.

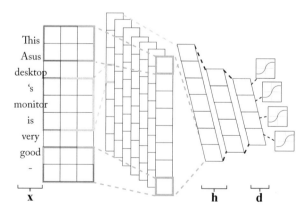

Figure 5.1: Overall network of DOC.

Traditional multi-class classifiers [Bendale and Boult, 2016, Goodfellow et al., 2016] typically use softmax as the final output layer, which does not have the rejection capability since the

[1]https://github.com/alexander-rakhlin/CNN-for-Sentence-Classification-in-Keras

probability of prediction for each class is normalized across all training/seen classes. Instead, a 1-vs.-rest layer is built containing N Sigmoid functions for N seen classes. For the i-th Sigmoid function corresponding to class l_i, DOC takes all examples with $y = l_i$ as positive examples and all the rest examples for $y \neq l_i$ as negative examples.

The model is trained with the objective of summation of all log loss of the N Sigmoid functions on the training data D:

$$\text{Loss} = \sum_{i=1}^{N} \sum_{j=1}^{n} -\mathbb{I}(y_j = l_i) \log p(y_j = l_i) \tag{5.5}$$
$$-\mathbb{I}(y_j \neq l_i) \log(1 - p(y_j = l_i)) \ ,$$

where $\mathbb{I}$ is the indicator function and $p(y_j = l_i) = \text{Sigmoid}(d_i^j)$ is the probability output from ith sigmoid function on the jth document's ith-dimension of $\boldsymbol{d}$.

During testing, we reinterpret the prediction of N Sigmoid functions to allow rejection, as shown in Equation (5.6). For the i-th Sigmoid function, we check if the predicted probability $\text{Sigmoid}(d_i)$ is less than a threshold t_i belonging to class l_i. If all predicted probabilities are less than their corresponding thresholds for a test example, the example is rejected; otherwise, its predicted class is the one with the highest probability:

$$\hat{y} = \begin{cases} reject, & \text{if } \text{Sigmoid}(d_i) < t_i, \forall l_i \in \mathcal{Y}; \\ \text{argmax}_{l_i \in \mathcal{Y}} \text{Sigmoid}(d_i), & \text{otherwise} \ . \end{cases} \tag{5.6}$$

When DOC was published, OpenMax [Bendale and Boult, 2016] was the state-of-the-art. It uses a classification network and add to it the rejection capability by utilizing the logits that are trained via the closed-world softmax function. One assumption of OpenMax is that examples with equally likely logits are more likely from the unseen or rejection class, which can be examples that are hard to classify. It also requires validation examples from the unseen/rejection class to tune the hyperparameters. In contrast, DOC uses the 1-vs.-rest sigmoid layer to provide a representation of all other classes (the rest of the seen classes and unseen classes), and to enable the 1 class to form a good boundary. Experimental results in Shu et al. [2017a] show that this basic DOC is already better than OpenMax. DOC is further improved by tightening the decision boundaries, which we discuss next.

5.3.2 REDUCING OPEN-SPACE RISK

Sigmoid function usually uses the default probability threshold of $t_i = 0.5$ for classification of each class i. But this threshold does not consider potential open space risks from unseen (rejection) class data. We can improve the boundary by increasing t_i. We use Figure 5.2 to illustrate. The x-axis represents d_i and y-axis is the predicted probability $p(y = l_i | d_i)$. The sigmoid function tries to push positive examples (belonging to the i-th class) and negative examples (belonging to the other seen classes) away from the y-axis via a high gain around $d_i = 0$, which serves as the default decision boundary for d_i with the probability threshold $t_i = 0.5$. As demonstrated by those three circles on the right-hand side of the y-axis, during testing, unseen class examples (circles) can easily fill in the gap between the y-axis and those dense positive (+) examples, which may reduce the recall of rejection and the precision of the i-th seen class prediction. Obviously, a better decision boundary is at $d_i = T$, where the decision boundary more closely "wrap" those dense positive examples with the probability threshold $t_i \gg 0.5$. Note that only t_i is used in classification decision making in this work and T is not used.

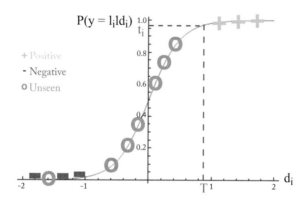

Figure 5.2: Open-space risk of sigmoid function and desired decision boundary $d_i = T$ and probability threshold t_i.

To obtain a better t_i for each seen class i-th, we use the idea of outlier detection in statistics.

1. Assume the predicted probabilities $p(y = l_i | x_j, y_j = l_i)$ of all training data of each class i follow one half of the Gaussian distribution (with mean $\mu_i = 1$), e.g., the three positive points in Figure 5.2 projected to the y-axis (we don't need d_i). We then artificially create the other half of the Gaussian distributed points (≥ 1): for each existing point $p(y = l_i | x_j, y_j = l_i)$, we create a mirror point $1 + (1 - p(y = l_i | x_j, y_j = l_i))$ (not a probability) mirrored on the mean of 1.

2. Estimate the standard deviation σ_i using both the existing points and the created points.

3. In statistics, if a value/point is a certain number (α) of standard deviations away from the mean, it is considered an outlier. We thus set the probability threshold $t_i = \max(0.5, 1 - \alpha\sigma_i)$. The commonly used number for α is 3, which also works well in our experiments.

Note that due to Gaussian fitting, different class l_i can have a different classification threshold t_i.

5.3.3   DOC FOR IMAGE CLASSIFICATION

DOC was originally proposed for open-world text classification. It was also later experimented for image classification and shown to perform very well [Shu et al., 2018], better than Open-Max [Bendale and Boult, 2016], which was designed for open image classification.

The evaluation used two publicly available image datasets: MNIST and EMNIST.

(1) **MNIST**[2]: MNIST is a well-known database of handwritten digits (10 classes), which has a training set of 60,000 examples, and a test set of 10,000 examples. In the experiment, 6 classes were used as the set of seen classes and the rest 4 classes were used as unseen classes.

(2) **EMNIST**[3] [Cohen et al., 2017]: EMNIST is an extension of MNIST to commonly used characters such as English alphabet. It is derived from the NIST Special Database 19. In the evaluation, EMNIST Balanced dataset with 47 balanced classes were used. It has a training set of 112,800 examples and a test set of 18,800 examples. 33 classes were used as the set of seen classes and another 10 classes were used as the unseen classes.

In Shu et al. [2018], DOC is compared with OpenMax [Bendale and Boult, 2016].[4] Both systems are based on deep learning. The results shown that DOC markedly outperforms OpenMax.

5.3.4   UNSEEN CLASS DISCOVERY

At the beginning of this chapter, we saw that in the second task of open-world learning a system or a human user identifies the hidden unseen classes in the rejected instances before they are incrementally learned in the third task. In Shu et al. [2018], an attempt was made to solve this problem automatically. In all previous works, this was done manually. The task is called *unseen class discovery*. The idea in Shu et al. [2018] is to transfer the class similarity knowledge learned from the seen classes to the hidden unseen classes. The transferred similarity knowledge is then used by a hierarchical clustering algorithm to cluster the rejected instances/examples to discover the hidden classes in the rejected instances. Note that this transfer of knowledge is from supervised learning to unsupervised learning.

This proposed transfer is warranted because we humans seem to group things based on our prior knowledge of what might be considered similar or different. For example, if we are given two objects and are asked whether they are of the same class/category or of different classes given some context, most probably we can tell. Why is that the case? We believe that we have learned in the past what are considered to be of the same class or of different classes in a knowledge context. The knowledge context here is important. For example, we have learned to recognize some breeds of dogs, which forms the knowledge context. When we are given two new/unseen breeds of dogs, we probably know that they are of different breeds. If we are given many different dogs from each of the two breeds, we probably can cluster them into two clusters. However, if

[2]http://yann.lecun.com/exdb/mnist/
[3]https://www.nist.gov/itl/iad/image-group/emnist-dataset
[4]https://github.com/abhijitbendale/OSDN

our previous knowledge only has classes such as dog, chicken, pig, cow, and sheep and we are given two different but unseen breeds of dogs, we probably will say that they are of the same kind/class and are dogs. However, if we are given a tiger and a rabbit, we will probably tell that they are from different classes.

To solve this problem, Shu et al. [2018] proposed a Pairwise Classification Network (PCN) to learn a binary classifier to predict whether two given examples are from the same class or different classes, i.e., $g(\mathbf{x}_p, \mathbf{x}_q)$. The positive training data of PCN consists of a set of pairs of intra-class (same class) examples, and the negative training data consists of a set of pairs of inter-class (different classes) examples all from seen classes. A hierarchical clustering method then uses the function $g(\mathbf{x}_p, \mathbf{x}_q)$ (which can be regarded as a distance/similarity function) to find the number of hidden classes (clusters) in the unseen/rejected class examples. Further details can be found in Shu et al. [2018].

5.4   SUMMARY AND EVALUATION DATASETS

As AI systems such as self-driving cars, mobile robots, chatbots, and personal intelligent assistants are increasingly working in real-life open environments and interacting with humans and/or automated systems, open-world learning is becoming increasingly important. An open-world learner should be able to detect new things that it has not seen before and learn them incrementally to become more and more knowledgeable. To some extent, we can regard such a learner as self-motivated because it actively identifies unseen objects and learns them to become more and more knowledge. Open-world learning is still highly challenging and needs a great deal of future research.

Although in this chapter we only discussed open-world learning in the supervised learning setting, it can be viewed from a broad perspective of detecting things unknown and learning the unknown things. It thus can be applied to any type of learning. For example, the learning method presented in Chapter 8, which continually spots and learns new knowledge in human-machine conversations, can also be seen as a form of open-world learning.

Fei et al. [2016] evaluated their method using the 100-products Amazon review dataset created by Chen and Liu [2014b][5] and the popular text classification dataset 20-Newsgroup.[6,7] The 100-products Amazon review dataset contains Amazon reviews from 100 different types of products. Each type of product (or domain) has 1,000 reviews. The 20-Newsgroup dataset contains news articles of 20 different topics. Each topic has about 1,000 articles. Shu et al. [2018] used image datasets MNIST[8] and EMNIST.[9] ImageNet[10] and its derivative datasets can also be used.

---

[5]https://www.cs.uic.edu/~zchen/downloads/KDD2014-Chen-Dataset.zip
[6]http://qwone.com/~jason/20Newsgroups/
[7]https://archive.ics.uci.edu/ml/datasets/Twenty+Newsgroups
[8]http://yann.lecun.com/exdb/mnist/
[9]https://www.nist.gov/itl/iad/image-group/emnist-dataset
[10]http://image-net.org/

CHAPTER 6

Lifelong Topic Modeling

Topic modeling has been used extensively to find topics in a large collection of text documents. A topic is a distribution over words. Those words with high probabilities in a topic indicate the topic. This set of words is often very useful in practice. Topic modeling is well suited for lifelong learning (LL) because topics learned in the past in related domains can be used to guide the model inference in the new or current domain [Chen and Liu, 2014a,b, Wang et al., 2016]. The *knowledge base* (KB) (Section 1.4) thus mainly stores the past topics. In this chapter, we use the terms *domain* and *task* interchangeably as in the existing research; each task is from a different domain. Even with the current simple LL techniques, lifelong topic modeling can already produce significantly better results than without LL regardless of whether the data (the text collection) is large or small. When the data size is small, LL is even more advantageous. For example, when the data is small, traditional topic models produce very poor results, but lifelong topic models can still generate very good topics. Ideally, as the KB expands, fewer modeling errors will incur. This is similar to our human learning. As we become more and more knowledgeable, it is easier for us to learn more and also less likely to make mistakes. In the following sections, we discuss several current representative techniques of lifelong topic modeling.

6.1 MAIN IDEAS OF LIFELONG TOPIC MODELING

Topic models, such as LDA (latent Dirichlet allocation) [Blei et al., 2003] and pLSA (Probabilistic latent semantic analysis) [Hofmann, 1999], are unsupervised learning methods for discovering topics from a set of text documents. They have been applied to numerous applications, e.g., opinion mining [Chen et al., 2014, Liu, 2012, Mukherjee and Liu, 2012, Zhao et al., 2010], machine translation [Eidelman et al., 2012], word sense disambiguation [Boyd-Graber et al., 2007], phrase extraction [Fei et al., 2014], and information retrieval [Wei and Croft, 2006]. In general, topic models assume that each document discusses a set of topics, probabilistically, a multinomial distribution over the set of topics, and each topic is indicated by a set of topical words, probabilistically, a multinomial distribution over the set of all words. The two kinds of distributions are called *document-topic distribution* and *topic-word distribution*, respectively. The intuition is that some words are more or less likely to be present given the topics of a document. For example, "sport" and "player" will appear more often in documents about sports; "rain" and "cloud" will appear more frequently in documents about weather.

However, fully unsupervised topic models tend to generate many inscrutable topics. The main reason is that the objective functions of topic models are not always consistent with human judgment [Chang et al., 2009]. To deal with this problem, we can use any of the following three approaches.

1. *Inventing better topic models*: This approach may work if a large number of documents is available. If the number of documents is small, regardless of how good the model is, it will not generate good topics simply because topic models are unsupervised learning methods and insufficient data cannot provide reliable statistics for modeling. Some form of supervision or external information beyond the given documents is necessary.

2. *Asking users to provide prior domain knowledge*: This approach asks the user or a domain expert to provide some prior domain knowledge. One form of knowledge can be in the form of *must-links* and *cannot-links*. A must-link states that two terms (or words) should belong to the same topic, e.g., *price* and *cost*. A cannot-link indicates that two terms should not be in the same topic, e.g., *price* and *picture*. Some existing *knowledge-based topic models* (e.g., Andrzejewski et al. [2009, 2011], Chen et al. [2013b,c], Hu et al. [2011], Jagarlamudi et al. [2012], Mukherjee and Liu [2012], Petterson et al. [2010], Xie et al. [2015]) have used such prior domain knowledge to produce better topics. However, asking the user to provide prior knowledge is problematic in practice because the user may not know what knowledge to provide and wants the system to discover useful knowledge for him/her. It also makes the approach non-automatic.

3. *Using lifelong topic modeling*: This approach incorporates LL in topic modeling. Instead of asking the user to provide prior knowledge, prior knowledge is learned and accumulated automatically in the modeling of previous tasks. For example, we can use the topics resulted from modeling of previous tasks as the prior knowledge to help the new task modeling. The approach works because of the observation that there are usually a great deal of sharing of concepts or topics across domains and tasks in natural language processing [Chen and Liu, 2014a,b], e.g., in sentiment analysis [Liu, 2012, 2015] as we discussed in the Preface of this book. We will give some examples shortly too.

We focus on the third approach. Following the definition in Chapter 1, each task here means to perform topic modeling on a set of documents of a particular domain. The KB stores all the topics obtained from each of the previous tasks, which are used in various ways as prior knowledge in different lifelong topic models.

At the beginning, the KB is either empty or filled with knowledge from an external source such as WordNet [Miller, 1995]. It grows with the results of incoming topic modeling tasks. Since all the tasks are about topic modeling, we use *domains* to distinguish the tasks. Two topic modeling tasks are different if their corpus domains are different. The scope of a domain is quite general. A domain can be a category (e.g., sports) or a product (e.g., camera) or an event (e.g., presidential election). We use $\mathcal{T}_1, \mathcal{T}_2, \ldots, \mathcal{T}_N$ to denote the sequence of previous tasks, $\mathcal{D}^p = \{\mathcal{D}_1, \mathcal{D}_2, \ldots, \mathcal{D}_N\}$ to denote their corresponding data or corpora, and use $\mathcal{T}_{N+1}$ to denote the new or current task with its data $\mathcal{D}_{N+1}$.

Key Questions in Lifelong Topic Modeling

For lifelong topic modeling to work, several questions need to be answered. Different models have different strategies to answer these questions.

1. What past knowledge should be retained and accumulated in the KB? As indicated above, in existing models, only the output topics from each previous domain/task are retained.

2. What kinds of knowledge should be used in the new domain modeling and how to mine such knowledge from the KB? Note that the raw past topics in the KB may not be directly used in topic modeling. Current lifelong topic models use must-link and cannot-link types of knowledge mined from the raw past topics stored in the KB.

3. How to assess the quality of knowledge and how to deal with possibly wrong knowledge? Previous modeling can make mistakes, and wrong knowledge from the past is often detrimental to new modeling.

4. How to apply the knowledge in the modeling process to generate better topics in the new domain?

Why Does Lifelong Topic Modeling Work?

The motivation for lifelong topic modeling is that topics from a large number of previous domains can provide high-quality knowledge to guide the modeling in the new domain to produce better topics. Although every domain is different, there is often a fair amount of concept or topic overlapping across domains. Using product reviews of different types of products (or domains) as an example, we observe that every product review domain probably has the topic of *price*, reviews of most electronic products share the topic of *battery*, and reviews of some products share the topic of *screen*. Topics produced from a single domain can be erroneous (i.e., a topic may contain some irrelevant words in its top ranked positions), but if a set of shared words among some topics generated from multiple domains can be found, these shared words are more likely to be correct or coherent for a particular topic. They can serve as a piece of prior knowledge to help topic modeling.

For example, we have product reviews from three domains. The classic topic model such as LDA [Blei et al., 2003] is used to generate a set of topics from each domain. Every domain has a topic about *price*, which is listed below with its top four words (words are ranked based on their probabilities under each topic).

- Domain 1: *price*, *color*, *cost*, *life*

- Domain 2: *cost*, *picture*, *price*, *expensive*

- Domain 3: *price*, *money*, *customer*, *expensive*

These topics are not perfect due to the incoherent words (words that do not indicate the main topic): *color*, *life*, *picture*, and *customer*. However, if we focus on those words that appear together in the same topic at least in two domains (the underlined words), we find the following two sets:

$$\{price, cost\} \text{ and } \{price, expensive\}.$$

The words in each of the sets are likely to belong to the same topic. As such, {*price*, *cost*} and {*price*, *expensive*} can serve as prior or past knowledge. That is, a piece of knowledge contains words that are semantically correlated. These two sets are called *must-links*.

   With the help of the knowledge, a new model can be designed to adjust the probability and improve the output topics for each of the above three domains or a new domain. Given the above knowledge indicating *price* and *cost* are related, *price* and *expensive* are related, a new topic may be found in Domain 1: *price, cost, expensive, color*, which has three coherent words in the top four positions rather than only two words as in the original topic. This represents a good topic improvement.

   In the next section, we review the LTM model [Chen and Liu, 2014a], which uses only must-links as prior knowledge. Its main idea is also applied in the LAST model for a sentiment analysis task [Wang et al., 2016]. In Section 6.3, we review the more advanced model AMC [Chen and Liu, 2014b], which can use both must-links and cannot-links as prior knowledge to model in a new domain with only a small set of documents. There is also another model called AKL (Automated Knowledge LDA) [Chen et al., 2014] that clusters past topics before mining must-links. Since both LTM and AMC improve AKL, AKL will not be discussed further.

6.2 LTM: A LIFELONG TOPIC MODEL

Lifelong Topic Model (LTM) was proposed in Chen and Liu [2014a]. It works in the following lifelong setting: At a particular point in time, a set of N previous modeling tasks have been performed. From each past task/domain data (or document set) $\mathcal{D}_i \in \mathcal{D}^p$, a set of topics $\mathcal{S}_i$ has been generated. Such topics are called *prior topics* (or *p-topics* for short). Topics from all past tasks are stored in the *Knowledge Base* (KB) $\mathcal{S}$ (known as the *topic base* in Chen and Liu [2014a]). At a new time point, a new task represented by a new domain document set $\mathcal{D}_{N+1}$ arrives for topic modeling. This is also called the *current domain*. LTM does not directly use the p-topics in $\mathcal{S}$ as knowledge to help its modeling. Instead, it mines *must-links* from $\mathcal{S}$ and uses the must-links as *prior knowledge* to help model inferencing for the $(N + 1)$th task. The process is dynamic and iterative. Once modeling on $\mathcal{D}_{N+1}$ is done, its resulting topics are added to $\mathcal{S}$ for future use. LTM has two key characteristics.

1. LTM's knowledge mining is targeted, meaning that it only mines useful knowledge from those relevant p-topics in $\mathcal{S}$. To do this, LTM performs a topic modeling on $\mathcal{D}_{N+1}$ first to find some initial topics and then uses these topics to find similar p-topics in $\mathcal{S}$. Those similar p-topics are used to mine must-links (knowledge) which are more likely to be

applicable and correct. These must-links are then used in the next iteration of modeling to guide the inference to generate more accurate topics.

2. LTM is a fault-tolerant model as it is able to deal with errors in automatically mined must-links. First, due to wrong topics (topics with many incoherent/wrong words or topics without a dominant semantic theme) in $\mathcal{S}$ or mining errors, the words in a must-link may not belong to the same topic in general. Second, the words in a must-link may belong to the same topic in some domains, but not in others due to the domain diversity. Thus, to apply such knowledge in modeling, the model must deal with possible errors in must-links.

6.2.1   LTM MODEL

Like many topic models, LTM uses Gibbs sampling for inference [Griffiths and Steyvers, 2004]. Its graphical model is the same as LDA, but it has a very different sampler which can incorporate prior knowledge and also handle errors in the knowledge as indicated above. The LTM system is illustrated in Figure 6.1, in the general LL framework of Figure 1.2.

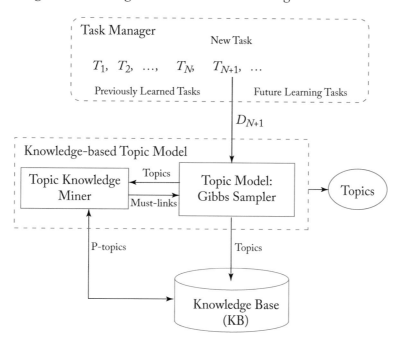

Figure 6.1: The Lifelong Topic Model (LTM) system architecture.

LTM works as follows (Algorithm 6.7): It first runs the **Gibbs sampler** of LTM for M iterations (or sweeps) to find a set of initial topics $\mathcal{A}_{N+1}$ from $\mathcal{D}_{N+1}$ with no knowledge (line 1). It then makes another M Gibbs sampling sweep (lines 2–5). But before each of these new sweeps, it first mines a set of targeted *must-links* (knowledge) $\mathcal{K}_{N+1}$ for every topic in $\mathcal{A}_{N+1}$

using the function **TopicKnowledgeMiner** (Algorithm 6.8, detailed in the next subsection) and then uses $\mathcal{K}_{N+1}$ to generate a new set of topics $\mathcal{A}_{N+1}$ from $\mathcal{D}_{N+1}$. To distinguish topics in $\mathcal{A}_{N+1}$ from p-topics, these new topics are called the *current topics* (or *c-topics* for short). We say that the mined must-links are targeted because they are mined based on the c-topics in $\mathcal{A}_{N+1}$ and are targeted at improving the topics in $\mathcal{A}_{N+1}$. Note that to make the algorithm more efficient, it is not necessary to mine knowledge for every sweep. Section 6.2.2 focuses on the topic knowledge mining function of LTM. The Gibbs sampler will be given in Section 6.2.4. Line 6 simply updates the knowledge base, which is simple, as each task is from a distinct domain in this paper. The set of topics is simply added to the **knowledge base $\mathcal{S}$** for future use.

Algorithm 6.7 Lifelong Topic Modeling (LTM)

Input: New domain data $\mathcal{D}_{N+1}$; Knowledge Base $\mathcal{S}$
Output: Topics from new domain $\mathcal{A}_{N+1}$

1: $\mathcal{A}_{N+1} \leftarrow$ GibbsSampler($\mathcal{D}_{N+1}, \emptyset, M$) // Run M iterations with no knowledge
2: **for** $i = 1$ **to** M **do**
3:    $\mathcal{K}_{N+1} \leftarrow$ TopicKnowledgeMiner($\mathcal{A}_{N+1}, \mathcal{S}$)
4:    $\mathcal{A}_{N+1} \leftarrow$ GibbsSampler($\mathcal{D}_{N+1}, \mathcal{K}_{N+1}, 1$) // Run with knowledge $\mathcal{K}_{N+1}$
5: **end for**
6: $\mathcal{S} \leftarrow$ UpdateKB($\mathcal{A}_{N+1}, \mathcal{S}$)

Algorithm 6.8 TopicKnowledgeMiner

Input: topics from new domain $\mathcal{A}_{N+1}$; knowledge base $\mathcal{S}$
Output: must-links (knowledge) $\mathcal{K}_{N+1}$ for new domain

---

1: **for** each p-topic $s_k \in \mathcal{S}$ **do**
2:    $j^* = \min_j$ KL-Divergence(a_j, s_k) for each c-topic $a_j \in \mathcal{A}_{N+1}$
3:    **if** KL-Divergence(a_{j^*}, s_k) $\leq \pi$ **then**
4:       $\mathcal{M}^{N+1}{}_{j^*} \leftarrow \mathcal{M}^{N+1}{}_{j^*} \cup \{s_k\}$
5:    **end if**
6: **end for**
7: $\mathcal{K}_{N+1} \leftarrow \cup_{j^*}$ FIM($\mathcal{M}^{N+1}{}_{j^*}$). // Frequent Itemset Mining

---

6.2.2   TOPIC KNOWLEDGE MINING

The **TopicKnowledgeMiner** function is given in Algorithm 6.8. For each p-topic s_k in $\mathcal{S}$, it finds the best matching (or the most similar) c-topic a_{j^*} in the c-topic set $\mathcal{A}_{N+1}$ (line 2). The matching is done using KL-Divergence (line 2) since each topic is a distribution over words. $\mathcal{M}^{N+1}{}_{j*}$ is used to store all matching p-topics for each c-topic a_{j*} (line 4). Note that the matching p-topics are found for each individual c-topic a_{j*} because a_{j*}-specific p-topics are preferable for more accurate knowledge (must-links) mining, which is done in line 7. In other words, these matching p-topics $\mathcal{M}^{N+1}{}_{j*}$ are targeted toward each a_{j*} and should provide high quality knowledge for a_{j*}. $\mathcal{M}^{N+1}{}_{j*}$ is mined to generate must-links $\mathcal{K}^{N+1}{}_{j*}$ for each c-topic a_{j*}. Must-links mined for all c-topics in $\mathcal{A}_{N+1}$ are stored in $\mathcal{K}_{N+1}$. Below, we describe topic matching and knowledge mining in greater detail.

Topic matching (lines 2–5, Algorithm 6.8): To find the best match for s_k in $\mathcal{S}$ with a c-topic a_{j*} in $\mathcal{A}_{N+1}$, KL-Divergence is used, which computes the difference between two distributions (lines 2 and 3). Specifically, Symmetrized KL (SKL) Divergence is employed, i.e., given two distributions P and Q, the divergence is calculated as:

$$SKL(P, Q) = \frac{KL(P, Q) + KL(Q, P)}{2} \quad , \text{ and} \tag{6.1}$$

$$KL(P, Q) = \sum_i \ln \left(\frac{P(i)}{Q(i)} \right) P(i) \quad . \tag{6.2}$$

The c-topic with the minimum SKL Divergence with s_k is denoted by a_{j*}. Parameter π is used to ensure that the p-topics in $\mathcal{M}^{N+1}{}_{j*}$ are reasonably correlated with a_{j*}.

Mining must-link knowledge using frequent itemset mining (FIM): Given the p-topics in each matching set $\mathcal{M}^{N+1}{}_{j*}$, this step finds sets of words that appear together multiple times in these p-topics. The shared words among matching p-topics across multiple domains are likely to belong to the same topic. To find such shared words in the matching set of p-topics $\mathcal{M}^{N+1}{}_{j*}$, frequent itemset mining (FIM) is used [Agrawal and Srikant, 1994].

FIM is stated as follows: Given a set of transactions $\mathcal{X}$, where each transaction $x_i \in \mathcal{X}$ is a set of items. In our context, x_i is a set of top words of a p-topic (no probability attached). $\mathcal{X}$ is $\mathcal{M}^{N+1}{}_{j*}$ without lowly ranked words in each p-topic as only the top words are usually representative of a topic. The goal of FIM is to find every itemset (a set of items) that satisfies some user-specified frequency threshold (also called *minimum support*), which is the minimum number of times that an itemset should appear in $\mathcal{X}$. Such itemsets are called *frequent itemsets*. In the context of LTM, a frequent itemset is a set of words that have appeared together multiple times in the p-topics of $\mathcal{M}^{N+1}{}_{j*}$, which is a must-link.

Only frequent itemsets of length two, i.e., each must-link has only two words, are used in the LTM model, e.g., {battery, life}, {battery, power}, {battery, charge}. Larger sets tend to contain more errors.

6.2.3   INCORPORATING PAST KNOWLEDGE

As each must-link reflects a possible semantic similarity relation between a pair of words, the *generalized Pólya urn* (GPU) model [Mahmoud, 2008] is used to leverage this knowledge in the

Gibbs sampler of LTM to encourage the pair of words to appear in the same topic. Below, we first introduce the Pólya urn model which serves as the basic framework to incorporate knowledge, and then presents the generalized Pólya urn model, which can deal with possible errors in must-links to make LTM fault-tolerant to some extent.

Simple Pólya Urn Model. The Pólya urn model works on colored balls and urns. In the topic model context, a term/word can be seen as a ball of a certain color and a topic as an urn. The distribution of a topic is reflected by the color proportions of balls in the urn. LDA follows the simple Pólya urn (SPU) model in the sense that when a ball of a particular color is drawn from an urn, the ball is put back to the urn along with a new ball of the same color. The content of the urn changes over time, which gives a self-reinforcing property known as "the rich get richer." This process corresponds to assigning a topic to a term in Gibbs sampling.

Generalized Pólya urn Model. The generalized Pólya urn (GPU) model [Chen and Liu, 2014a, Mahmoud, 2008, Mimno et al., 2011] differs from SPU in that, when a ball of a certain color is drawn, two balls of that color are put back along with a certain number of balls of some other colors. These additional balls of some other colors added to the urn increase their proportions in the urn. This is the key technique for incorporating must-links as we will see below.

Applying the GPU model to topic modeling, when a word w is assigned to a topic t, each word w' that shares a must-link with w is also assigned to the topic t by a certain amount, which is decided by the matrix $\mathbf{A}'_{t,w',w}$. w' is thus promoted by w, meaning that the probability of w' under topic t is also increased. Here, a must-link of a topic t means this must-link is extracted from the p-topics matching with topic t.

The problem is how to set proper values for matrix $\mathbf{A}'_{t,w',w}$. To answer this question, let us also consider the problem of wrong knowledge. Since the must-links are mined from p-topics in multiple previous domains automatically, the semantic relationship of words in a must-link may not be correct for the current domain. It is a challenge to determine which must-link is not appropriate. One way to deal with the problem is to assess how the words in a must-link correlated with each other in the current domain. If they are more correlated, they are more likely to be correct for a topic in the domain and thus should be promoted more. If they are less correlated, they are more likely to be wrong and should be promoted less (or even not promoted).

To measure the correlation of two words in a must-link in the current domain, Pointwise Mutual Information (PMI) is used, which is a measure of word association in text [Church and Hanks, 1990]. In this case, it measures the extent to which two words tend to co-occur, which corresponds to "the higher-order co-occurrence" on which topic models are based [Heinrich, 2009]. The PMI of two words is defined as follows:

$$PMI(w_1, w_2) = \log \frac{P(w_1, w_2)}{P(w_1)P(w_2)} \quad , \tag{6.3}$$

where $P(w)$ denotes the probability of seeing word w in a random document, and $P(w_1, w_2)$ denotes the probability of seeing both words co-occurring in a random document. These probabilities are empirically estimated using the current domain collection $\mathcal{D}_{N+1}$:

$$P(w) = \frac{\#\mathcal{D}_{N+1}(w)}{\#\mathcal{D}_{N+1}} , \qquad (6.4)$$

$$P(w_1, w_2) = \frac{\#\mathcal{D}_{N+1}(w_1, w_2)}{\#\mathcal{D}_{N+1}} , \qquad (6.5)$$

where $\#\mathcal{D}_{N+1}(w)$ is the number of documents in $\mathcal{D}_{N+1}$ that contain word w, and $\#\mathcal{D}_{N+1}(w_1, w_2)$ is the number of documents that contain both words w_1 and w_2. $\#\mathcal{D}_{N+1}$ is the total number of documents in $\mathcal{D}_{N+1}$. A positive PMI value implies a true semantic correlation of words, while a non-positive PMI value indicates little or no semantic correlation. Thus, only must-links with positive PMI values are considered. A parameter factor μ is added to control how much the GPU model should trust the word relationships indicated by PMI. The amount of promotion for word w' when seen w is defined as follows:

$$\mathbf{A}'_{t,w,w'} = \begin{cases} 1 & w = w' \\ \mu \times PMI(w, w') & (w, w') \text{ is a must-link of topic } t \\ 0 & \text{otherwise} . \end{cases} \qquad (6.6)$$

6.2.4   CONDITIONAL DISTRIBUTION OF GIBBS SAMPLER

The GPU model is nonexchangeable, i.e., the joint probability of the words in any given topic is not invariant to the permutation of those words. The inference for the model can be computationally expensive due to the nonexchangeability of words, that is, the sampling distribution for the word of interest depends on each possible value for the subsequent words along with their topic assignments. LTM takes the approach of Mimno et al. [2011] which approximates the true Gibbs sampling distribution by treating each word as if it were the last. The approximate Gibbs sampler has the following conditional distribution:

$$\begin{aligned} P(z_i = t | \mathbf{z}^{-i}, \mathbf{w}, \alpha, \beta, \mathbf{A}') \propto \\ \frac{n_{d,t}^{-i} + \alpha}{\sum_{t'=1}^{T}(n_{d,t'}^{-i} + \alpha)} \times \frac{\sum_{w'=1}^{V} \mathbf{A}'_{t,w',w_i} \times n_{t,w'}^{-i} + \beta}{\sum_{v=1}^{V}(\sum_{w'=1}^{V} \mathbf{A}'_{t,w',v} \times n_{t,w'}^{-i} + \beta)} , \end{aligned} \qquad (6.7)$$

where n^{-i} is the count excluding the current assignment of z_i, i.e., z^{-i}, $\mathbf{w}$ refers to all the words in all documents in the document collection $\mathcal{D}_{N+1}$, and w_i is the current word to be sampled with a topic denoted by z_i. $n_{d,t}$ denotes the number of times that topic t was assigned to words in document d, where d is the document index of word w_i. $n_{t,v}$ refers to the number of times that word v appears under topic t. α and β are predefined Dirichlet hyperparameters. T is the number of topics, and V is the vocabulary size. $\mathbf{A}'$ is the promotion matrix defined in Equation (6.6).

6.3   AMC: A LIFELONG TOPIC MODEL FOR SMALL DATA

The LTM model needs a fairly large set of documents in order to generate reasonable initial topics to be used in finding similar past topics in the knowledge base to mine appropriate must-link knowledge. However, when the document set (or data) is very small, this approach does not work because the initial modeling produces very poor topics, which cannot be used to find matching or similar past topics in the knowledge base to serve as prior knowledge. A new approach is thus needed. The AMC model (topic modeling with Automatically generated Must-links and Cannot-links) [Chen and Liu, 2014b] aims to solve the problem. AMC's must-link knowledge mining does not use any information from the new domain/task. Instead, it mines must-links from the past topics independent of the new domain. However, to make the resulting topics accurate, must-link knowledge is far from sufficient. Thus, AMC also uses cannot-links, which are hard to mine independent of the new domain data due to the high computational complexity. Cannot-links are mined dynamically. All these are detailed in this section.

Algorithm 6.9 AMC Model

Input: New domain data $\mathcal{D}_{N+1}$; Knowledge Base $\mathcal{S}$
Output: Topics from new domain $\mathcal{A}_{N+1}$

1: $\mathcal{M} \leftarrow$ MustLinkMiner($\mathcal{S}$)
2: $\mathcal{C} = \emptyset$ // $\mathcal{C}$ stores cannot-links
3: $\mathcal{A}_{N+1} \leftarrow$ GibbsSampler($\mathcal{D}_{N+1}, \mathcal{M}, \mathcal{C}, M$); // Run M Gibbs iterations with must-links $\mathcal{M}$ but no cannot-links
4: **for** $r = 1$ **to** R **do**
5:      $\mathcal{C} \leftarrow \mathcal{C} \cup$ CannotLinkMiner($\mathcal{S}, \mathcal{A}_{N+1}$)
6:      $\mathcal{A}_{N+1} \leftarrow$ GibbsSampler($\mathcal{D}_{N+1}, \mathcal{M}, \mathcal{C}, N$)
7: **end for**
8: $\mathcal{S} \leftarrow$ UpdateKB($\mathcal{A}_{N+1}, \mathcal{S}$)

6.3.1   OVERALL ALGORITHM OF AMC

Algorithm 6.9 gives the overall algorithm of AMC, which is also illustrated in Figure 6.2. Line 1 mines a set of must-links $\mathcal{M}$ using the function **MustLinkMiner** from previous topics (or p-topics) in the **knowledge base** (KB) $\mathcal{S}$. Note here the must-links can be generated offline independent of the current new task. Line 3 runs the proposed **Gibbs sampler** (introduced in Section 6.3.5) using only the must-links $\mathcal{M}$ to produce a set of topics $\mathcal{A}_{N+1}$, where M is the number of Gibbs sampling iterations. Line 5 mines cannot-links $\mathcal{C}$ using the function **CannotLinkMiner** based on the current topics $\mathcal{A}_{N+1}$ and p-topics in the knowledge base $\mathcal{S}$ (see Section 6.3.3). Then line 6 uses both must-links $\mathcal{M}$ and cannot-links $\mathcal{C}$ to improve the resulting topics. This process can run iteratively (R times) to obtain a set of superior topics to be stored in the knowledge base and also output to the user. Function $UpdateKB(\mathcal{A}_{N+1}, \mathcal{S})$ (line 8) is simple at the moment. If the domain of $\mathcal{A}_{N+1}$ exists in $\mathcal{S}$, replace those topics of the domain in $\mathcal{S}$ with $\mathcal{A}_{N+1}$; otherwise, $\mathcal{A}_{N+1}$ is added to $\mathcal{S}$.

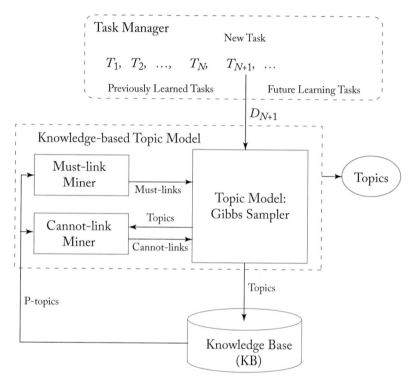

Figure 6.2: The AMC model system architecture.

6.3.2 MINING MUST-LINK KNOWLEDGE

Since AMC cannot use topics from the new domain to find similar topics in the knowledge base (KB) like LTM, it mines must-links directly from the KB without considering the context of any new task using the function **MustLinkMiner**. Recall that each topic generated from a topic model, such as LDA, is a distribution over words, i.e., words with their associated probabilities. Words are commonly ranked based on their probabilities in a descending order. In practice, top words under a topic are expected to represent some similar semantic meaning. The lower ranked words usually have very low probabilities due to the smoothing effect of the Dirichlet hyperparameters rather than true correlations within the topic, leading to their unreliability. Thus, in Chen and Liu [2014b], only the top 15 words are employed to represent a topic. This topic representation is used in mining both must-link and cannot-link knowledge.

Given knowledge base $\mathcal{S}$, similar to the LTM model in Section 6.2, must-links are sets of words that appear together in multiple topics and they are mined using the data mining technique *frequent itemset mining* (FIM). However, this technique is insufficient due to the problem with the single minimum support threshold used in classic FIM algorithms.

A single minimum support is not appropriate because generic topics, such as *price* with topic words like *price* and *cost*, are shared by many (even all) product review domains, but specific topics, such as *screen*, occur only in product domains having such features. This means that different topics may have very different frequencies in the data. Thus, using a single minimum support threshold is unable to extract both generic and specific topics because if this threshold is set too low, the popular topics will result in numerous spurious frequent itemsets (which results in wrong must-links) and if it is set too high, must-links from less frequent topics will not be found. This is called the *rare item problem* in data mining and has been well documented in the data mining literature [Liu, 2007].

To address the above issue, the AMC model uses the *multiple minimum supports frequent itemset mining* (MS-FIM) algorithm in Liu et al. [1999]. MS-FIM is stated as follows: Given a set of transactions $\mathcal{R}$, where each transaction $r_i \in \mathcal{R}$ is a set of items from a global item set $\mathcal{I}$, i.e., $r_i \subseteq \mathcal{I}$. In AMC, r_i is a topic comprising the top words of the topic (no probability attached). An item is a word. $\mathcal{R}$ is thus the collection of all p-topics in the knowledge base $\mathcal{S}$ and $\mathcal{I}$ is the set of all words in $\mathcal{S}$. In MS-FIM, each item/word is given a minimum itemset support (MIS). The minimum support that an itemset (a set of items) must satisfy is not fixed. It depends on the MIS values of all the items in the itemset. MS-FIM also has another constraint, called the *support difference constraint* (SDC), expressing the requirement that the supports of the items in an itemset must not be too different. MIS and SDC together can solve the above rare item problem.

The goal of MS-FIM is to find all itemsets that satisfy the user-specified MIS thresholds and SDC constraints. Such itemsets are called *frequent itemsets*. In AMC, a frequent itemset is a set of words which have appeared multiple times in the p-topics in the knowledge base $\mathcal{S}$. The frequent itemsets of length two are used as the learned must-link knowledge, e.g., {battery, life}, {battery, power}, {battery, charge}, {price, expensive}, {price, pricy}, {cheap, expensive}.

Then again each must-link used in AMC has only two words [Chen and Liu, 2014b]. As mentioned in Section 6.2.2, larger sets tend to contain more errors. Such errors are hard to deal with than those in pairs. The same rationale applies to cannot-links.

There are two key challenges in incorporating the must-link knowledge in modeling.

1. A word can have multiple meanings or senses. For example, *light* may mean "something that makes things visible" or "of little weight." Different senses may lead to distinct must-links. For example, with the first sense of *light*, the must-links can be {light, bright} and {light, luminance}. In contrast, {light, weight} and {light, heavy} indicate the second sense of light. Without dealing with this, it can cause the transitivity problem [Chen and Liu, 2014b]. That is, if words w_1 and w_2 form a must-link, and words w_2 and w_3 form a must-

link, it implies a must-link between w_1 and w_3, i.e., w_1, w_2, and w_3 should be in the same topic. With transitivity, *light*, *bright*, and *weight* would be incorrectly assumed to be in the same topic.

2. Not every must-link is suitable for a domain. This is the same wrong knowledge problem discussed in Section 6.2.

To deal with the first issue, a must-link graph was proposed in Chen and Liu [2014b] to distinguish multiple senses in must-links to solve the transitivity problem. As the must-links are automatically mined from the set of p-topics (for past topics) in $\mathcal{S}$, the p-topics may also provide some guidance on whether the mined must-links share the same word sense or not. Given two must-links m_1 and m_2, if they share the same word sense, the p-topics that cover m_1 should have some overlapping with p-topics that cover m_2. For example, must-links {light, bright} and {light, luminance} should be mostly coming from the same set of p-topics related to the semantic meaning "something that makes things visible" of *light*. On the other hand, little topic overlapping indicates likely different word senses. For example, must-links {light, bright} and {light, weight} may come from two different sets of p-topics as they usually refer to different topics.

Following this idea, a must-link graph G is constructed where each must-link is a vertex. An edge is formed between two vertices if the two must-links m_1 and m_2 have a shared word. For each edge, the amount of their original p-topics overlapping is used to decide whether the two must-links share the same sense or not. Given two must-links m_1 and m_2, the p-topics in $\mathcal{S}$ covering each of them are denoted by $\mathcal{C}_1$ and $\mathcal{C}_2$, respectively. m_1 and m_2 share the same sense if

$$\frac{\#(\mathcal{C}_1 \cap \mathcal{C}_2)}{Max(\#\mathcal{C}_1, \#\mathcal{C}_2)} > \pi_{overlap} \quad , \tag{6.8}$$

where $\pi_{overlap}$ is the *overlap threshold* for distinguishing senses. This threshold is necessary due to errors in topics. The edges that do not satisfy the above inequality are discarded. The final must-link graph G can provide some guidance on selecting the right must-links sharing the same word sense in the Gibbs sampler (in Section 6.3.5) for dealing with the transitivity problem.

To tackle the second problem, Pointwise Mutual Information (PMI) was also used to approximate the semantic correlation using the current domain data. This is similar to that in the LTM model (Section 6.2.3) and thus will not be discussed again.

6.3.3 MINING CANNOT-LINK KNOWLEDGE

Although it is reasonable to find must-links from all past topics, it is problematic to find cannot-links from all past topics (p-topics) as it is prohibitive to do so. This is because for a word w, there are usually only a few words w_m that share must-links with w while there is a huge number of words w_c that can form cannot-links with w. In general, if there are V words in the vocabulary of all tasks or domains, then there are $O(V^2)$ potential cannot-links. However, for a new domain

$\mathcal{D}_{N+1}$, most of these cannot-links are not useful because the vocabulary size of $\mathcal{D}_{N+1}$ is much smaller than V. Thus, AMC focuses only on those words that are relevant to $\mathcal{D}_{N+1}$.

Formally, given the knowledge base $\mathcal{S}$ and the current c-topics $\mathcal{A}_{N+1}$ from the new task domain data $\mathcal{D}_{N+1}$, cannot-links from each pair of top words w_1 and w_2 in each c-topic $A_j \in \mathcal{A}_{N+1}$ are extracted. Based on this formulation, to mine cannot-links (using **CannotLinkMiner**), the mining algorithm enumerates every pair of top words w_1 and w_2 and checks whether they form a cannot-link or not. Thus, the cannot-link mining is targeted to each c-topic with the aim to improve the c-topic using the discovered cannot-links.

Given two words, CannotLinkMiner determines whether they form a cannot-link or not as follows: If the words seldom appear together in p-topics in $\mathcal{S}$, they are likely to have distinct semantic meanings. Let the number of past domains that w_1 and w_2 appear in different p-topics be N_{diff} and the number of past domains that w_1 and w_2 share the same topic be N_{share}. N_{diff} should be much larger than N_{share}. Two conditions or thresholds are necessary to control the formation of a cannot-link:

1. The ratio $N_{diff}/(N_{share} + N_{diff})$ (called the *support ratio*) is equal to or larger than a threshold π_c. This condition is intuitive.

2. N_{diff} is greater than a support threshold π_{diff}. This condition is needed because the above ratio can be 1, but N_{diff} can be very small and thus unreliable.

Some cannot-link examples are as follows: {battery, money}, {life, movie}, {battery, line} {price, digital}, {money, slow}, and {expensive, simple}.

Similar to must-links, cannot-links can be wrong too. Like must-links, there are also two cases: (a) A cannot-link contains terms that have semantic correlations. For example, {battery, charger} is not a correct cannot-link. (b) A cannot-link does not fit for a particular domain. For example, {card, bill} is a correct cannot-link in the camera domain, but not appropriate for restaurants. Wrong cannot-links are usually harder to detect and to verify than wrong must-links. Due to the power-law distribution of natural language words [Zipf, 1932], most words are rare and will not co-occur with most other words. The low co-occurrences of two words do not necessarily mean a negative correlation (cannot-link). Chen and Liu [2014b] proposed to detect and balance cannot-links inside the sampling process. More specifically, they extended the Pólya urn model to incorporate the cannot-link knowledge, and also to deal with the issues above.

6.3.4 EXTENDED PÓLYA URN MODEL

Gibbs sampler for the AMC model differs from that of LTM as LTM does not consider cannot-links. A *multi-generalized Pólya Urn* (M-GPU) model was proposed in Chen and Liu [2014b] for AMC. We have introduced the simple Pólya urn (SPU) model, and the generalized Pólya urn (GPU) model in Section 6.2.3. We now extend the GPU model to the *multi-generalized Pólya urn model* (M-GPU).

Instead of involving only one urn at a time as in the SPU and GPU models, the M-GPU model considers a set of urns in the sampling process simultaneously [Chen and Liu, 2014b]. M-GPU allows a ball to be transferred from one urn to another, enabling multi-urn interactions. Thus, during sampling, the populations of several urns will evolve even if only one ball is drawn from one urn. This capability makes the M-GPU model more powerful and suitable for solving the complex problems discussed so far.

In M-GPU, when a ball is randomly drawn, a certain number of additional balls of each color are returned to the urn, rather than just two balls of the same color as in SPU. This is inherited from the GPU model. As a result, the proportions of these colored balls are increased, making them more likely to be drawn in this urn in the future. This is called the *promotion* of these colored balls in Chen and Liu [2014b]. Applying the idea, when a word w is assigned to a topic k, each word w' that shares a must-link with w is also assigned to topic k by a certain amount $\lambda_{w',w}$. The definition of $\lambda_{w',w}$ is similar to the promotion matrix in the LTM model (see Section 6.2.3). Thus, we will not discuss it further here.

To deal with multiple senses problem in M-GPU, Chen and Liu [2014b] exploited the fact that each word usually has only one correct sense or meaning under one topic. Since the semantic concept of a topic is usually represented by some top words under it, the word sense that is the most related to the concept is treated as the correct sense. If a word w does not have multiple must-links, then there is no multiple sense problem. If w has multiple must-links, the rationale here is to sample a must-link (say m) that contains w to be used to represent the likely word sense from the must-link graph G. The sampling distribution will be given in the next sub-section. Then, the must-links that share the same word sense with m, including m, are used to promote the related words of w.

To deal with cannot-links, M-GPU defines two sets of urns to be used in sampling. The first set is the set of topic urns $U_{d \in \mathcal{D}_{N+1}}^K$, where each urn is for one document and contains balls of K colors (topics) and each ball inside has a color $k \in \{1 \dots K\}$. This corresponds to the document-topic distribution in AMC. The second set of urns is the set of word urns $U_{k \in \{1 \dots K\}}^V$ corresponding to the topic-word distributions, with balls of colors (words) $w \in \{1 \dots V\}$ in each word urn.

Based on the definition of cannot-link, two words in a cannot-link cannot both have large probabilities under the same topic. As M-GPU allows multi-urn interactions, when sampling a ball representing word w from a word urn U_k^V, the balls representing the cannot-words of w, say w_c (sharing cannot-links with w) can be transferred to other urns (see step 5 below), i.e., decreasing the probabilities of those cannot-words (words in a cannot-link) under this topic while increasing their corresponding probabilities under some other topic. The ball representing word w_c should be transferred to an urn which has a higher proportion of w_c. That is, an urn that has a higher proportion of w_c is randomly sampled for w_c to transfer to (step 5b below). However, it is possible that there is no other urn that has a higher proportion of w_c. There are two ways to deal with this issue. (1) Create a new urn to move w_c to, which was used in Chen et al.

[2013c]. This approach assumes that the cannot-link is correct. (2) Keep w_c in the urn U_k^V as the cannot-link may not be correct, so it is possible that U_k^V is the right urn for w_c. As discussed in Section 6.3.3, a cannot-link can be wrong. For example, the model puts *battery* and *life* in the same topic k where *battery* and *life* have the highest probabilities (or proportions). However, a cannot-link {battery, life} wants to separate them after seeing them in the same topic. In this case, we should not trust the cannot-link as it wants to split the correlated words into different topics. Chen and Liu [2014b] took the second approach due to the noise in cannot-links.

Based on all the above ideas, the M-GPU sampling scheme is presented as follows.

1. Sample a topic k from U_d^K and a word w from U_k^V sequentially, where d is the dth document in $\mathcal{D}_{N+1}$.

2. Record k and w, put back two balls of color k into urn U_d^K, and two balls of color w into urn U_k^V.

3. Sample a must-link m that contains w from the prior knowledge base. Get a set of must-links $\{m'\}$ where m' is either m or a neighbor of m in the must-link graph G.

4. For each must-link $\{w, w'\}$ in $\{m'\}$, we put back $\lambda_{w',w}$ number of balls of color w' into urn U_k^V based on matrix $\lambda_{w',w}$.

5. For each word w_c that shares a cannot-link with w:

   (a) Draw a ball q_c of color w_c (to be transferred) from U_k^V and remove it from U_k^V. The document of ball q_c is denoted by d_c. If no ball of color w_c can be drawn (i.e., there is no ball of color w_c in U_k^V), skip steps (b) and (c).

   (b) Produce an urn set $\{U_{k'}^V\}$ such that each urn in it satisfies the following conditions:
   (i) $k' \neq k$,
   (ii) The proportion of balls of color w_c in $U_{k'}^V$ is higher than that of balls of color w_c in U_k^V.

   (c) If $\{U_{k'}^V\}$ is not empty, randomly select one urn $U_{k'}^V$ from it. Put the ball q_c drawn from Step a) into $U_{k'}^V$. Also, remove a ball of color k from urn $U_{d_c}^K$ and put back a ball of k' into urn $U_{d_c}^K$. If $\{U_{k'}^V\}$ is empty, put the ball q_c back to U_k^V.

6.3.5   SAMPLING DISTRIBUTIONS IN GIBBS SAMPLER

For each word w_i in each document d, sampling consists of two phases based on the M-GPU sampling process above.

Phase 1 (steps 1–4 in M-GPU): calculate the conditional probability of sampling a topic for word w_i. The process enumerates each topic k and calculates its corresponding probability, which is decided by three sub-steps.

(a) Sample a must-link m_i that contains w_i, which is likely to have the word sense consistent with topic k, based on the following conditional distribution:

$$P(m_i = m|k) \propto P(w_1|k) \times P(w_2|k) ,\qquad (6.9)$$

where w_1 and w_2 are the words in must-link m and one of them is the same as w_i. $P(w|k)$ is the probability of word w under topic k given the current status of the Markov chain in the Gibbs sampler, which is defined as:

$$P(w|k) \propto \frac{\sum_{w'=1}^{V} \lambda_{w',w} \times n_{k,w'} + \beta}{\sum_{v=1}^{V} (\sum_{w'=1}^{V} \lambda_{w',v} \times n_{k,w'} + \beta)} ,\qquad (6.10)$$

where $n_{k,w}$ refers to the number of times that word w appears under topic k. β is the predefined Dirichlet hyper-parameter.

(b) After getting the sampled must-link m_i, a set of must-links $\{m'\}$ are created where m' is either m_i or a neighbor of m_i in the must-link graph G. The must-links in this set $\{m'\}$ are likely to share the same word sense of word w_i according to the corresponding edges in the must-link graph G.

(c) The conditional probability of assigning topic k to word w_i is defined as below:

$$\begin{aligned}
p(z_i &= k|z^{-i}, \boldsymbol{w}, \alpha, \beta, \lambda) \\
&\propto \frac{n_{d,k}^{-i} + \alpha}{\sum_{k'=1}^{K} (n_{d,k'}^{-i} + \alpha)} \\
&\times \frac{\sum_{\{w',w_i\}\in\{m'\}} \lambda_{w',w_i} \times n_{k,w'}^{-i} + \beta}{\sum_{v=1}^{V} (\sum_{\{w',v\}\in\{m_v'\}} \lambda_{w',v} \times n_{k,w'}^{-i} + \beta)} ,
\end{aligned}\qquad (6.11)$$

where n^{-i} is the count excluding the current assignment of z_i, i.e., z^{-i}. $\boldsymbol{w}$ refers to all the words in all documents in the new document collection $\mathcal{D}_{N+1}$, and w_i is the current word to be sampled with a topic denoted by z_i. $n_{d,k}$ denotes the number of times that topic k is assigned to words in document d. $n_{k,w}$ refers to the number of times that word w appears under topic k. α and β are predefined Dirichlet hyper-parameters. K is the number of topics, and V is the vocabulary size. $\{m_v'\}$ is the set of must-links sampled for each word v following Phase 1 (a) and (b), which is recorded during the iterations.

Phase 2 (step 5 in M-GPU): this sampling phase deals with cannot-links. There are two sub-steps.

(a) For every cannot-word (say w_c) of w_i, one instance (say q_c) of w_c from topic z_i is sampled, where z_i denotes the topic assigned to word w_i in Phase 1, based on the following conditional distribution:

$$P(q = q_c|\boldsymbol{z}, \boldsymbol{w}, \alpha) \propto \frac{n_{d_c,k} + \alpha}{\sum_{k'=1}^{K} (n_{d_c,k'} + \alpha)} ,\qquad (6.12)$$

where d_c denotes the document of the instance q_c. If there is no instance of w_c in z_i, skip step b).

(b) For each drawn instance q_c from Phase 2 (a), resample a topic k (not equal to z_i) based on the conditional distribution below:

$$
\begin{aligned}
&P(z_{q_c} = k | \boldsymbol{z}^{-q_c}, \boldsymbol{w}, \alpha, \beta, \lambda, q = q_c) \\
&\propto \boldsymbol{I}_{[0, p(w_c | k)]}(P(w_c | z_c)) \\
&\times \frac{n_{d_c,k}^{-q_c} + \alpha}{\sum_{k'=1}^{K}(n_{d_c,k'}^{-q_c} + \alpha)} \\
&\times \frac{\sum_{\{w', w_c\} \in \{m_c'\}} \lambda_{w', w_c} \times n_{k, w'}^{-q_c} + \beta}{\sum_{v=1}^{V}(\sum_{\{w', v\} \in \{m_v'\}} \lambda_{w', v} \times n_{k, w'}^{-q_c} + \beta)},
\end{aligned}
\tag{6.13}
$$

where z_c (the same as z_i sampled from (6.11)) is the original topic assignment. $\{m_c'\}$ is the set of must-links sampled for word w_c. Superscript $-q_c$ denotes the counts excluding the original assignments. $\boldsymbol{I}()$ is an indicator function, which restricts the ball to be transferred only to an urn that contains a higher proportion of word w_c. If there is no topic k that has a higher proportion of w_c than z_c, then keep the original topic assignment, i.e., assign z_c to w_c.

6.4 SUMMARY AND EVALUATION DATASETS

Although lifelong supervised learning (LSL) has been researched since the beginning of LL at around 1995, little research had been done on lifelong unsupervised learning until recently. Topic modeling is an unsupervised learning method. Several papers were published in the past few years on lifelong topic modeling. These methods all exploit the sharing of topics and concepts across tasks and domains in natural language. As discussed earlier in chapter 1, natural language processing (NLP) is quite suitable for LL precisely due to its extensive sharing of expressions, concepts, and syntactic structures across domains and tasks. We thus believe LL can have a major impact on NLP.

Here we would also like to highlight a question that people often ask about lifelong un-supervised learning. That is, when faced with a new task, can we combine all the past and the current data to form a big dataset to perform the task to achieve the same or even better results for the new task? This combining data approach can be seen as a very simple form of LL. How-ever, this approach is not suitable for lifelong topic modeling because of three key reasons. First, with a large number of different domain datasets, there will be a huge number of topics, which makes it very difficult for the user to set the number of topics. Second, much poorer topics are likely to be the result due to the mix-up of the data from very different domains which cause wrong words to be grouped together to form incoherent topics. Thus, true topics specific to the new domain may be lost or mixed up with topics from other domains. Third, because of the fact that the new data is only a tiny portion of the big data, topic modeling will not focus on those small and domain-specific topics but only on those big topics that cut cross many domains. Then those important domain-specific topics will be lost.

We now list some evaluation datasets, which are mainly constructed from product reviews. Chen and Liu [2014a] created a dataset containing online reviews from 50 domains (types of products), which are all electronic products. The reviews were crawled from Amazon.com. Each domain has 1,000 reviews. This dataset has been used in Chen and Liu [2014a] and Wang et al. [2016]. This dataset also has four larger review collections with 10,000 reviews in each collection. The dataset is publicly available.[1] Chen and Liu [2014b] expanded this dataset by adding another 50 domains of reviews, each of which contains reviews from a non-electronic product or domain. Some example product domains include Bike, Tent, Sandal, and Mattress. Again, each domain contains 1,000 reviews. This larger dataset is also available publicly.[2]

---

[1]https://www.cs.uic.edu/~zchen/downloads/ICML2014-Chen-Dataset.zip
[2]https://www.cs.uic.edu/~zchen/downloads/KDD2014-Chen-Dataset.zip

CHAPTER 7

Lifelong Information Extraction

This chapter focuses on *lifelong information extraction.* Information extraction (IE) is a rich area for applying lifelong learning (LL) as the goal of IE is to continually extract and accumulate useful information or knowledge as much as possible. In other words, the extraction process is by nature continuous and cumulative. The extracted information earlier can be used to help extract more information later with higher quality [Carlson et al., 2010a, Liu et al., 2016, Shu et al., 2017b]. These all match the goal of LL. In this case, the knowledge base (KB) of LL often stores the extracted information and some other forms of useful knowledge.

The most well-known lifelong information extraction system is NELL, which stands for *Never-Ending Language Learner* [Carlson et al., 2010a, Mitchell et al., 2015]. NELL is the only lifelong semi-supervised learning system that we are aware of. NELL is also a good example of the systems approach to LL. It is perhaps the only live LL system that has been reading the Web to extract certain types of information (or knowledge) 24 hours a day and 7 days a week since January 2010. Although several efforts have been made by other researchers to read the Web to extract various types of knowledge to build large KBs, e.g., WebKB [Craven et al., 1998], KnowItAll [Etzioni et al., 2004], and YAGO [Suchanek et al., 2007], they are not LL systems, except ALICE [Banko and Etzioni, 2007]. ALICE works in an LL setting and is unsupervised. Its goal is to extract information to build a domain theory of concepts and their relationships. The extraction in ALICE is done using a set of handcrafted lexico-syntactic patterns (e.g., "< ? grains > such as buckwheat" and "buckwheat is a < ? food >"). ALICE also has some ability to produce *general propositions* by *abstraction*, which deduces a general proposition from a set of extracted fact instances. ALICE's LL feature is realized by updating the current domain theory with new extractions and by using the output of each learning cycle to suggest the focus of subsequent learning tasks, i.e., the process is guided by earlier learned knowledge. This chapter focuses on NELL and some more recent lifelong IE techniques such as AER [Liu et al., 2016] and L-CRF [Shu et al., 2017b].

7.1   NELL: A NEVER-ENDING LANGUAGE LEARNER

A large part of human knowledge is gained by reading books and listening to lectures. Unfortunately, computers still cannot understand human language in order to read books to acquire knowledge systematically. The NELL system represents an effort to extract two types of knowledge from reading Web documents. Since January 2010, it has been reading the Web non-stop and has accumulated millions of facts with attached confidence weights (e.g., servedWith(tea, biscuits)), which are called *beliefs*, and are stored in a structured knowledge base.

NELL is a lifelong semi-supervised information extraction system, and it has only a small number of labeled training examples for each of its learning tasks, which is far from sufficient to learn accurate extractors to extract reliable knowledge. Without reliable knowledge, LL is impossible because using wrong knowledge for future learning is highly detrimental. As we discussed several times earlier in the book, identifying the correct past knowledge is a major challenge for LL. NELL made an attempt to solve this problem by extracting different types of related knowledge using different types of data sources and by constraining the learning tasks so that the tasks can reinforce or help each other and constrain each other to ensure each of them extracts reasonably correct or robust knowledge.

The *input* to NELL consists of the following:

1. an ontology defining a set of target categories and relations to be learned (in the form of a collection of *predicates*), a handful of seed training examples for each, and a set of constraints that couple various categories and relations (e.g., Person and Sport are mutually exclusive);

2. webpages crawled from the Web, which NELL uses to extract information; and

3. occasional interactions with human trainers to correct some of the mistakes made by NELL.

With this input, the *goal* of NELL is two-fold.

1. Extract facts from the webpages to populate the initial ontology. Specifically, NELL continuously extracts the following two types of information or knowledge.

    (a) *category* of a noun or noun phrase, e.g., Los Angeles is a *city*, Canada is a *country*, New York Yankees is a *baseball team*.[1]

    (b) *relations* of a pair of noun phrases. For example, given the name of a university (say Stanford) and a major (say Computer Science), check whether the relation hasMajor(Stanford, Computer Science) is true.

2. Learn to perform the above extraction tasks, also called the reading tasks, better than yesterday. Learning is done in a semi-supervised manner.

To achieve these objectives, NELL works iteratively in an infinite loop, i.e., hence *never-ending* or *lifelong*, like an EM algorithm. Each iteration performs two main tasks corresponding to the two objectives:

---

[1]Recently learned knowledge examples in NELL can be found at http://rtw.ml.cmu.edu/rtw/.

1. *Reading task*: read and extract the two types of information or knowledge from the Web to grow the KB of structured facts (or beliefs). Specifically, NELL's category and relation extractors first propose extraction results as updates to the KB. The Knowledge Integrator (KI) module then records these individual recommendations, makes the final decision about the confidence assigned to each potential belief after considering various consistency constraints, and then performs updates to the KB.

   Because of a huge number of possible candidate beliefs and the large size of the KB, NELL considers only the beliefs in which it has the highest confidences, limiting each extractor or sub-system to propose only a limited number of new candidate beliefs for any given predicate on any given iteration. This enables NELL to operate tractably and also to be able to add millions of new beliefs over many iterations.

2. *Learning task*: learn to read better with the help of the accumulated knowledge in the up-dated KB and coupling constraints. The evidence for improved reading is shown by the fact that the system can extract more information more accurately. Specifically, learning in NELL optimizes the accuracy of each learned function. The training examples consist of a combination of human-labeled instances (the dozen or so labeled seed examples pro-vided for each category and relation in NELL's ontology), labeled examples contributed over time through NELL's crowdsourcing website, a set of NELL self-labeled training examples corresponding to NELL's current/updated knowledge base, and a large amount of unlabeled Web text. The last two sets of the training examples propel NELL's LL and self-improvement process over time.

   Since semi-supervised learning often gives low accuracy because of the limited number of labeled examples, NELL improves the accuracy and the quality of the extracted knowledge by coupling the simultaneous training of many extractors. These extractors extract from different data sources and are learned using different learning algorithms. The rationale is that the errors made by these extractors are uncorrelated. When multiple subsystems make uncorrelated errors, the probability that they all make the same error is much lower, which is the product of individual probabilities (considering them as independent events). These multiple extractors are linked by coupling constraints. That is, the under-constrained semi-supervised learning tasks can be made more robust by imposing constraints that arise from coupling the training of many extractors for different categories and relations. Their learning tasks are thus guided by one another's results, through the shared KB and coupling constraints.

Even with coupling constraints and sophisticated mechanisms to ensure extraction quality, errors can still be made, which may propagate, accumulate, and even multiply. NELL mitigates this problem further by interacting with some human trainer each day for about 10–15 minutes to fix some of the errors to prevent their propagation and producing poorer and poorer results subsequently.

7.1.1   NELL ARCHITECTURE

NELL's architecture is shown in Figure 7.1. There are four main components in NELL: data resources, knowledge base, subsystem components, and knowledge integrator.

Data Resources. Since the goal of NELL is to continuously read webpages crawled from the Web to extract knowledge, webpages are thus the data resources.

Knowledge Base. Knowledge Base (KB) stores all the extracted knowledge that is expressed as beliefs. As mentioned above, two types of knowledge are stored in the knowledge base: instances of various categories and relations. A piece of knowledge can be a candidate fact or a belief. A candidate fact is extracted and proposed by the subsystem components, and may be promoted to a belief, which is decided by Knowledge Integrator.

Subsystem Components. It contains several subsystems, which are the extractors and learning components of NELL. As indicated earlier, in the reading phrase, these subsystems perform extraction and propose candidate facts to be included in the knowledge base. In the learning phrase, they learn based on their individual learning methods with the goal of improving themselves using the current state of the knowledge base and the coupling constraints. Each subsystem is built based on different extraction methods taking input from different parts of the data resources. The four subsystems, CPL, CSEAL, CMC, and RL, are discussed in the next section.

Knowledge Integrator. Knowledge Integrator (KI) controls the condition of promoting candidate facts into beliefs. It consists of a set of hand-coded criteria. Specifically, KI decides what candidate facts are promoted to the status of beliefs. It is based on a hard-coded rule. The rule says that candidate facts with high confidence from a single source (those with posterior > 0.9) are promoted. Low-confidence candidates are promoted if they have been proposed by multiple sources. Mutual-exclusion and type-checking constraints are also used in KI. In particular, if a candidate fact does not satisfy a constraint (mutual-exclusion or type-checking) based on the existing beliefs in the KB, it is not promoted. Once a candidate fact becomes a belief, it never gets demoted.

7.1.2   EXTRACTORS AND LEARNING IN NELL

As we can see in Figure 7.1, there are four major subsystem components that perform extraction and learning [Carlson et al., 2010a]. We discuss them now.

- *Coupled Pattern Learner* (CPL): In the reading phrase, the extractors in the CPL subsystem extract both category and relation instances from unstructured free Web text using contextual patterns. At the beginning, they are the given seed patterns and later they are the learned and promoted patterns from previous iterations. Example category and relation extraction patterns are "mayor of X" and "X plays for Y", respectively.

  In the learning phrase, such patterns are learned in CPL using some heuristics procedures-based co-occurrence statistics between noun phrases and existing contextual patterns (both

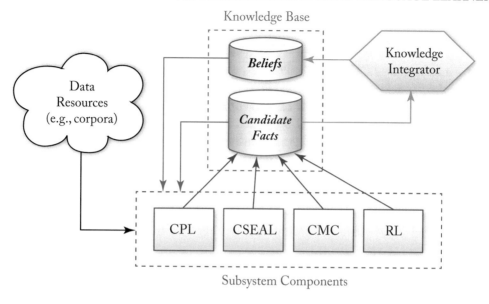

Figure 7.1: **NELL** system architecture [Carlson et al., 2010a].

defined using part-of-speech tag sequences) for each predicate of interest. The learned patterns essentially serve as classification functions that classify noun phrases by semantic categories [Mitchell et al., 2015] (e.g., a Boolean-valued function that classifies whether any given noun phrase refers to a city). Relationships between predicates are used to filter out patterns that are too general.

Candidate instances and patterns extracted and learned are also filtered using mutual-exclusion and type-checking constraints to remove those possible invalid instances and patterns. Mutual-exclusion constraints enforce that mutually exclusive predicates cannot both be satisfied by the same input x. For example, x cannot be both a person and a car. Type-checking constraints are used to couple or to link relation extractors (or contextual patterns for relation extraction) with category extractors (or contextual patterns for category extraction). For example, given the relation universityHasMajor(x, y), x should be of type/category *university* and y should be of type/category *major*. Otherwise, the relation is likely to be wrong.

The remaining candidates are then ranked using simple co-occurrence statistics and estimated precisions. Only a small number of top-ranked candidate instances and patterns are promoted and retained in the knowledge base for future use. Additional details about CPL can be found in Carlson et al. [2010b].

- *Coupled SEAL* (CSEAL): CSEAL is an extraction and learning system that extracts facts from semi-structured webpages using wrapper induction. Its core system is an existing wrapper induction system called SEAL [Wang and Cohen, 2009]. SEAL is based on an semi-supervised ML model called *set-expansion*, also known as *learning from positive and unlabeled examples* (*PU learning*) [Liu et al., 2002]. Set expansion or PU learning is defined as follows. Given a set S of seeds of a certain target type (or positive examples), and a set of unlabeled examples U (which is obtained by querying the Web using the seeds), the goal of set-expansion is to identify examples in U that belong to S. SEAL uses *wrappers*. For a category, its wrapper is defined by character strings which specify the left context and right context of an entity to be extracted. The entities are mined from Web lists and tables of the category. For example, a wrapper <li class="player arg1"> <h4><a href="#"> indicates that *arg1* should be a player. An instance is extracted by a wrapper if it is found anywhere in the document matching with the left and right contexts of the wrapper. The relations are extracted in the same manner. However, wrappers for these predicates are learned independently in SEAL. SEAL does not have the mechanism for exploiting mutual exclusion or type-checking constraints. CSEAL added these constraints on top of SEAL so that the candidates extracted from the wrappers can be filtered out if they violate the mutual-exclusion and type-checking constraints.

  Again, the remaining candidates are ranked and only a small number of top-ranked candidate instances and patterns are promoted and retained in the KB for future use. Additional details about CSEAL can be found in Carlson et al. [2010b] and Wang and Cohen [2009].

- *Coupled Morphological Classifier* (CMC): CMC consists of a set of binary classifiers, one for each category, for classifying whether the extracted candidate facts/beliefs by other components/subsystems are indeed of their respective categories. To ensure high precision, the system classifies only up to 30 new beliefs per predicate in each iteration, with a minimum posterior probability of 0.75. All classifiers are built using logistic regression with L_2 regularization. The features are various morphological clues, such as words, capitalization, affixes, and parts-of-speech. The positive training examples are obtained from the beliefs in the current KB, and negative examples are items inferred using mutual-exclusion constraints and the current beliefs in the KB.

- *Rule Learner* (RuleL): RuleL is a first-order relational learning system similar to FOIL [Quinlan and Cameron-Jones, 1993]. Its goal is to learn probabilistic Horn clauses and to use them to infer new relations from the existing relations in the KB. This reasoning capability represents an important advance of NELL that does not exist in most current information extraction or LL systems.

In Mitchell et al. [2015], several new subsystem components were also proposed, e.g., NEIL (Never Ending Image Learner), which classifies a noun phrase using its associated visual images, and OpenEval (an online information validation technique), which uses real-time Web search to gather the distribution of text contexts around a noun phrase to extract instances of predicates. More information about them and others can be found in the original paper.

7.1.3   COUPLING CONSTRAINTS IN NELL

We have already seen two types of coupling constraints, i.e., mutual-exclusion and type-checking constraints. NELL also uses several other coupling constraints to ensure the quality or precision of its extraction and learning results. We believe that coupling constraints are an important feature and novelty of NELL, which help solve a key problem in LL, i.e., how to ensure that the learned or extracted knowledge is correct (see Section 1.4). Without a reasonable solution to this problem, LL is difficult because errors can propagate and even multiply as the iterative process progresses. Below are three other coupling constraints that NELL uses.

- *Multi-view co-training coupling constraint*: In many cases, the same category or relation can be learned from different data sources, or *views*. For example, a predicate instance can be learned from free text by CPL and also extracted from some semi-structured webpages by CSEAL using its wrapper. This constraint requires that the two results should agree with each other. In general, for extraction or learning categories, given a noun phrase X, multiple functions that use different sets of noun phrase features (or views) to predict if X belongs to a category Y_i should give the same result. The same idea also applies to extraction or learning of relations.

- *Subset/superset coupling constraint*: When a new category is added to NELL's ontology, its parents (supersets) are also specified. For example, "Snack" is declared to be a subset of "Food." If X belongs to "Snack," then X should satisfy the constraint of being "Food." This constraint couples or links the learning tasks that extract "Snack" to those that learn to extract "Food."

- *Horn clause coupling constraint*: The probabilistic Horn clauses learned from FOIL [Quinlan and Cameron-Jones, 1993] give another set of logic-based constraints. For example, X living in Chicago and Chicago being a city in the U.S. can lead to X lives in U.S. (with a probability p). In general, whenever NELL learns a Horn clause rule to infer new beliefs from existing beliefs in the KB, this rule serves as a coupling constraint.

7.2   LIFELONG OPINION TARGET EXTRACTION

This section introduces an application of LL to a specific unsupervised IE task based on the work in Liu et al. [2016]. The IE task is aspect or opinion target extraction from opinion documents, which is a fundamental task in sentiment analysis [Liu, 2012]. It aims to extract opinion targets from opinion text. For example, from the sentence "*This phone has a great screen, but its battery life is short*," it should extract "screen" and "battery life." In product reviews, aspects are product attributes or features.

In Liu et al. [2016], a syntactic dependency-based method called *double propagation* (DP) [Qiu et al., 2011] was adopted as the base extraction method, which was augmented with the LL capability. DP is based on the fact that opinions have targets and there are often syntactic relations between sentiment or opinion words (e.g., "great" in *"the picture quality is great"*) and target aspect (e.g., "picture quality"). Due to the syntactic relations, sentiment words can be recognized by identified aspects, and aspects can be identified by known sentiment words. The extracted sentiment words and aspects are used to identify new sentiment words and new aspects, which are used again to extract more sentiment words and aspects. This bootstrapping propagation process ends when no more sentiment words or aspects can be found. The extraction rules were designed based on dependency relations among sentiment words and aspects produced by dependency parsing.

Figure 7.2 shows the dependency relations between words in *"The phone has a good screen."* If "good" is a known sentiment word (given or extracted), "screen," a noun modified by "good," is an aspect as they have a dependency relation *amod*. From a given seed set of sentiment words, one can extract a set of aspects from a syntactic rule like *"if a word A, whose part-of-speech (POS) is a singular noun (nn), has the dependency relation amod with (i.e., modified by) a sentiment word O, then A is an aspect."* Similarly, one can use such rules to extract new aspects from the extracted aspects, and new sentiment words from the extracted aspects.

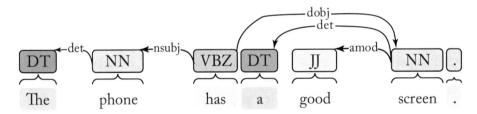

Figure 7.2: Dependency relations in the sentence *"The phone has a good screen."*

7.2.1   LIFELONG LEARNING THROUGH RECOMMENDATION

Although effective, syntactic rule-based methods such as DP still have room for major improvements. Liu et al. [2016] showed that incorporating LL can improve the extraction markedly.

To realize LL, Liu et al. [2016] used the idea of recommendation, in particular *collaborative filtering* [Adomavicius and Tuzhilin, 2005]. This type of recommendation uses the behavioral information of other users to recommend products/services to the current user. Following the idea, Liu et al. [2016] used the information in reviews of a large number of other products (data of the previous tasks) to help extract aspects from reviews of the current product (data of the new task). The recommendation is based on the previous task data and extraction results. This method is called *lifelong IE through recommendations*. Two forms of recommendations were used: (1) *semantic similarity-based recommendation* and (2) *aspect associations-based recommendation*.

1. Semantic similarity-based recommendation aims to solve the problem of missing synonymous aspects of DP using word vectors trained from a large corpus of reviews for similarity comparison. Word vectors are regarded as a form of prior or past knowledge learned from the past data. Let us see an example. Using the DP method, "picture" is extracted as an aspect from the sentence "*The picture is blurry,*" but "photo" is not extracted from the sentence "*The phone is good, but not the photos.*" One reason for the inability to extract "photo" is that to ensure good extraction precision, many useful syntactic dependency rules with low precision are not used. The proposed semantic similarity-based recommendation can make use of the extracted aspect "picture" to recommend "photo" ("photo" is a synonym of "picture") based on the semantic similarity of the word vectors of the two words.

2. The second form of recommendation is via aspect associations or correlations. This form is useful because in the first recommendation, "picture" cannot be used to recommend "battery" as an aspect because their semantic similarity value is very small. The idea of using the second form of recommendation is that many aspects are correlated or co-occur across domains. For example, those products with the aspect "picture" also have a high chance of using batteries, as pictures are usually taken by digital devices that need batteries. If rules about such associations can be discovered, they can be used to recommend additional aspects. For this purpose, association rules from data mining [Liu, 2007] were employed. To mine associations, Liu et al. [2016] used the extraction results from the previous tasks stored in the knowledge base $\mathcal{S}$.

The knowledge base contains two forms of information: the word vectors and the extraction results from the previous/past tasks.

7.2.2 AER ALGORITHM

The proposed extraction algorithm is called AER (Aspect Extraction based on Recommendations) [Liu et al., 2016] and is shown in Algorithm 7.10, which consists of three main steps: base extraction, aspect recommendation (which includes a knowledge learning sub-step discussed in Section 7.2.3), and KB updating.

Step 1 (base extraction, lines 1–2): Given the new document data $\mathcal{D}_{N+1}$ for the $(N+1)$th task for extraction and a set $\mathcal{O}$ of seed opinion or sentiment words, this step first uses the DP method (DPextract) to extract an initial (or base) set $\mathcal{A}^-$ of aspects using a set $\mathcal{R}^-$ of high precision rules (line 1). The set of high precision rules are selected from the set of rules in DP by evaluating their precisions individually using a development set. The set $\mathcal{A}^-$ of extracted aspects thus has very high precision but not high recall. Then, it extracts a set $\mathcal{A}^+$ of aspects from a larger set $\mathcal{R}^+$ of high recall rules ($\mathcal{R}^- \subseteq \mathcal{R}^+$) also using DPextract (line 2). The set $\mathcal{A}^+$ of extracted aspects thus has very high recall but low precision.

Step 2 (aspect recommendation, lines 3–7): This step recommends more aspects using $\mathcal{A}^-$ as the base to improve the recall. To ensure recommendation quality, Liu et al. [2016]

required that the recommended aspects must be from the set $\mathcal{A}^{diff} = \mathcal{A}^+ - \mathcal{A}^-$ (line 3). As indicated above, two forms of recommendation are performed, similarity-based using Sim-recom (line 4) and association rule-based using AR-recom (line 6). Their respective results $\mathcal{A}^s$ and $\mathcal{A}^a$ are combined with $\mathcal{A}^-$ to produce the final extraction result (line 7). Note that the word vectors $\mathcal{WV}$ required by Sim-recom are stored in the knowledge base $\mathcal{S}$. The association rules $\mathcal{AR}$ used in AR-recom are mined from the extraction results of previous tasks also stored in the knowledge base $\mathcal{S}$.

Step 3 (knowledge base update, line 8): This step updates the knowledge base $\mathcal{S}$, which is simple, as each task is from a distinct domain in this paper [Liu et al., 2016]. That is, the set of extracted aspects is simply added to the knowledge base $\mathcal{S}$ for future use.

We will not discuss step 1 and step 3 further as they are fairly straightforward. Our focus is on the two recommendation methods, which will be introduced in Section 7.2.4. For the recommendations to work, we first need to learn the past knowledge in terms of word vectors $\mathcal{WV}$ and association rules $\mathcal{AR}$. We discuss knowledge learning next.

Algorithm 7.10 AER Algorithm

Input: New domain data $\mathcal{D}_{N+1}$, high precision aspect extraction rules $\mathcal{R}^-$, high recall aspect extraction rules $\mathcal{R}^+$, seed opinion words $\mathcal{O}$, and knowledge base $\mathcal{S}$
Output: Extracted aspect set $\mathcal{A}$

1: $\mathcal{A}^- \leftarrow$ DPextract($\mathcal{D}_{N+1}, \mathcal{R}^-, \mathcal{O}$) // $\mathcal{A}^-$: aspect set with high precision
2: $\mathcal{A}^+ \leftarrow$ DPextract($\mathcal{D}_{N+1}, \mathcal{R}^+, \mathcal{O}$) // $\mathcal{A}^+$: aspect set with high recall
3: $\mathcal{A}^{diff} \leftarrow \mathcal{A}^+ - \mathcal{A}^-$
4: $\mathcal{A}^s \leftarrow$ Sim-recom($\mathcal{A}^-, \mathcal{A}^{diff}, \mathcal{WV}$) // $\mathcal{WV}$ is the set of word vectors stored in the knowledge base $\mathcal{S}$
5: $\mathcal{AR} \leftarrow$ MineAssociationRules($\mathcal{S}$)
6: $\mathcal{A}^a \leftarrow$ AR-recom($\mathcal{A}^-, \mathcal{A}^{diff}, \mathcal{AR}$)
7: $\mathcal{A} \leftarrow \mathcal{A}^- \cup \mathcal{A}^s \cup \mathcal{A}^a$
8: $\mathcal{A} \leftarrow$ UpdateKB($\mathcal{A}, \mathcal{S}$)

7.2.3 KNOWLEDGE LEARNING

Generating Word Vectors

In Liu et al. [2016], word vectors were trained using neural networks in Mikolov et al. [2013b]. Researchers have shown that using word vectors trained this way is highly effective for the purpose of semantic similarity comparison [Mikolov et al., 2013b, Turian et al., 2010]. There are several publicly available word vector resources trained from Wikipedia, Reuters News, or Broadcast News for general NLP tasks such as POS tagging and Named Entity Recognition [Collobert and Weston, 2008, Huang et al., 2012, Pennington et al., 2014]. However, the initial experiments in Liu et al. [2016] showed these word vectors were not accurate for their task. They thus trained the word vectors using a large corpus of 5.8 million reviews [Jindal and Liu, 2008]. Clearly, the word vectors can also be trained by just using the data in the past domains, but the paper did not try that. It will be interesting to see the difference in results produced with word vectors trained from the two data sources. Note that using word vectors in extraction can be regarded as a simple form of LL because the generation of word vectors basically uses the data from previous tasks to learn a rich representation (word vector) of each word to be used in the current extraction task. As the system sees more data, the word vectors can also be updated.

Mining Association Rules
Association rules are of the form, $X \longrightarrow Y$, where X and Y are disjoint sets of items, i.e., a set of aspects in our case. X and Y are called *antecedent* and *consequent* of the rule, respectively. The *support* of the rule is the number of transactions that contains both X and Y divided by the total number of transactions, and the *confidence* of the rule is the number of transactions that contains both X and Y divided by the number of transactions that contains X. Given a transaction database *DB*, an association rule-mining algorithm generates all rules that satisfy a user-specified *minimum support* and a *minimum confidence* constraint [Agrawal and Srikant, 1994]. DB contains a set of transactions. In our case, a transaction consists of all the aspects discovered from one previous domain or task, which is stored in the knowledge base $\mathcal{S}$. Association rule mining has been well studied in data mining.

7.2.4   RECOMMENDATION USING PAST KNOWLEDGE

Recommending Aspects using Word Vectors
Algorithm 7.11 gives the details of Sim-recom($\mathcal{A}^-$, $\mathcal{A}^{diff}$, $\mathcal{WV}$), which recommends aspects based on aspect similarities measured using word vectors. For each term t in $\mathcal{A}^{diff}$, which can be a single word or a multi-word phrase, if the similarity between t and any term in $\mathcal{A}^-$ is at least ϵ (line 2), which means that t is very likely to be an aspect and should be recommended, then add t into $\mathcal{A}^s$ (line 3). The final recommended aspect set is $\mathcal{A}^s$.

The function Sim(t, $\mathcal{A}^-$) in line 2 returns the maximum similarity between term t and the set of terms in $\mathcal{A}^-$, i.e.,

$$\text{Sim}(t, \mathcal{A}^-) = \max\{\text{VS}(\boldsymbol{\phi}_t, \boldsymbol{\phi}_{t_q}) : t_q \in \mathcal{A}^-\} , \tag{7.1}$$

where $\boldsymbol{\phi}_t$ is t's vector, $\text{VS}(\boldsymbol{\phi}_t, \boldsymbol{\phi}_{t_q})$ is $\text{VS}^w(\boldsymbol{\phi}_t, \boldsymbol{\phi}_{t_q})$ if t and t_q are single words, otherwise, $\text{VS}(\boldsymbol{\phi}_t, \boldsymbol{\phi}_{t_q})$ is $\text{VS}^p(\boldsymbol{\phi}_t, \boldsymbol{\phi}_{t_q})$. $\text{VS}^w(\boldsymbol{\phi}_t, \boldsymbol{\phi}_{t_q})$ and $\text{VS}^p(\boldsymbol{\phi}_t, \boldsymbol{\phi}_{t_q})$ compute single words similarity and phrases or phrase-word similarity, respectively. Given two terms t and t', their semantic similarity is calculated using their vectors $\boldsymbol{\phi}_t$ and $\boldsymbol{\phi}_{t'}$ in $\mathcal{WV}$ as below:

$$\text{VS}^w(\boldsymbol{\phi}_t, \boldsymbol{\phi}_{t'}) = \frac{\boldsymbol{\phi}_t^T \boldsymbol{\phi}_{t'}}{||\boldsymbol{\phi}_t|| \cdot ||\boldsymbol{\phi}_{t'}||} . \tag{7.2}$$

Algorithm 7.11 Sim-recom Algorithm

Input: Aspect sets $\mathcal{A}^-$ and $\mathcal{A}^{diff}$, word vectors $\mathcal{WV}$
Output: Recommended aspect set $\mathcal{A}^s$

1: **for** each aspect term $t \in \mathcal{A}^{diff}$ **do**
2:    **if** $\text{Sim}(t, \mathcal{A}^-) \geq \epsilon$ **then**
3:        $\mathcal{A}^s \leftarrow \mathcal{A}^s \cup \{t\}$
4:    **end if**
5: **end for**

Since there are no vectors for multi-word phrases in the pre-trained word vectors, the average cosine similarities of words in the phrases is used to evaluate phrase similarities:

$$\text{VS}^p(\boldsymbol{\phi}_t, \boldsymbol{\phi}_{t'}) = \frac{\sum_{i=1}^{L} \sum_{j=1}^{L'} \text{VS}^w(\boldsymbol{\phi}_{t_i}, \boldsymbol{\phi}_{t'_j})}{L \times L'} , \tag{7.3}$$

where L is the number of single words in t, and L' is that of t'. The reason for using average similarity for multi-word phrases is that it considers the length of the phrases, and sets lower similarity value naturally if the lengths of two phrases are different.

Recommending Aspects using Association Rules
Algorithm 7.12 gives the details of AR-recom, which recommends aspects based on aspect association rules. For each association rule r in $\mathcal{AR}$, if the antecedent of r is a subset of $\mathcal{A}^-$ (line 2), then recommend the terms in $\text{cons}(r) \cap \mathcal{A}^{diff}$ into the set $\mathcal{A}^a$ (line 3). The function $\text{ante}(r)$ returns the set of aspects in r's antecedent, and the function $\text{cons}(r)$ returns the set of (candidate) aspects in r's consequent.

Algorithm 7.12 AR-recom Algorithm

Input: Aspect sets $\mathcal{A}^-$ and $\mathcal{A}^{diff}$, association rules $\mathcal{AR}$
Output: Recommended aspect set $\mathcal{A}^a$

1: **for** each association rule $r \in \mathcal{AR}$ **do**
2:    **if** $\text{ante}(r) \subseteq \mathcal{A}^-$ **then**
3:        $\mathcal{A}^a \leftarrow \mathcal{A}^a \cup (\text{cons}(r) \cap \mathcal{A}^{diff})$
4:    **end if**
5: **end for**

For example, one association rule in $\mathcal{AR}$ could be: *picture, display* $\rightarrow$ *video, purchase*, whose antecedent contains "picture" and "display," and consequent contains "video" and "purchase." If both words "picture" and "display" are in $\mathcal{A}^-$, and only "video" is in $\mathcal{A}^{diff}$, then only "video" is added into $\mathcal{A}^a$.

7.3   LEARNING ON THE JOB

It is known that about 70% of our human knowledge comes from "on-the-job" learning. Only about 10% is learned through formal training and the rest 20% is learned through observation of others. For a machine learner to learn on the job or learn while working, it must continuously learn after model training. This section describes a simple method that does a limited form of learning on the job in the context of information extraction [Shu et al., 2017b]. Specifically, the paper shows that if the system has performed extraction from many (past) domains and retained their results as knowledge, Conditional Random Fields (CRF) [Lafferty et al., 2001] can leverage this knowledge in an LL manner to extract in a new domain better than the traditional CRF without using this prior knowledge. The proposed method is called L-CRF (*lifelong CRF*), and it is applied to aspect (product feature/attribute) extraction in sentiment analysis.

The main idea of L-CRF is that even after supervised model training, the model can still improve its extraction in testing or application. The improvement is possible because of a fair amount of aspect sharing across domains. Such sharing can be leveraged to help CRF perform better for new domains.

The setting of L-CRF is as follows: A CRF model M has been trained with a labeled training online review dataset. At a particular point in time, M has extracted aspects from data in N previous domains $D_1, \ldots, D_N$ (which are unlabeled) and the extracted sets of aspects are $A_1, \ldots, A_N$. Now, the system is faced with a new domain data D_{N+1}. M can leverage some *reliable prior knowledge* in $A_1, \ldots, A_N$ to make a better extraction from D_{N+1} than without leveraging the prior knowledge.

7.3.1   CONDITIONAL RANDOM FIELDS

CRF learns from an observation sequence $\mathbf{x}$ to estimate a label sequence $\mathbf{y}$: $p(\mathbf{y}|\mathbf{x}; \boldsymbol{\theta})$, where $\boldsymbol{\theta}$ is a set of weights. Let l be the l-th position in the sequence. The core parts of CRF are a set of feature functions $\mathcal{F} = \{f_h(y_l, y_{l-1}, \mathbf{x}_l)\}_{h=1}^{H}$ and their corresponding weights $\boldsymbol{\theta} = \{\theta_h\}_{h=1}^{H}$.

Feature Functions: Two types of feature functions (FF) are used in Shu et al. [2017b]. One is *Label–Label (L2)* FF:

$$f_{ij}^{LL}(y_l, y_{l-1}) = \mathbb{1}\{y_l = i\}\mathbb{1}\{y_{l-1} = j\}, \forall i, j \in \mathcal{Y} \;, \tag{7.4}$$

where $\mathcal{Y}$ is the set of labels, and $\mathbb{1}\{\cdot\}$ an indicator function. The other is *Label–Word (LW)* FF:

$$f_{iv}^{LW}(y_l, \mathbf{x}_l) = \mathbb{1}\{y_l = i\}\mathbb{1}\{\mathbf{x}_l = v\}, \forall i \in \mathcal{Y}, \forall v \in \mathcal{V} \;, \tag{7.5}$$

where $\mathcal{V}$ is the vocabulary. This FF returns 1 when the l-th word is v and the l-th label is v's specific label i; otherwise 0. $\mathbf{x}_l$ is the current word, and is represented as a multi-dimensional vector. Each dimension in the vector is a feature of $\mathbf{x}_l$.

The feature set {W, -1W, +1W, P, -1P, +1P, G} is used, where W is the word and P is its POS-tag, -1W is the previous word, -1P is its POS-tag, +1W is the next word, +1P is its POS-tag, and G is the generalized dependency feature.

Under the Label-Word FF type, two sub-types of FF are employed: *Label-dimension* FF and *Label-G* FF. Label-dimension FF is for the first six features, and Label-G is for the G feature.

The *Label-dimension (Ld)* FF is defined as

$$f_{iv^d}^{Ld}(y_l, \mathbf{x}_l) = \mathbb{1}\{y_l = i\}\mathbb{1}\{\mathbf{x}_l^d = v^d\}, \forall i \in \mathcal{Y} \forall v^d \in \mathcal{V}^d \ , \tag{7.6}$$

where $\mathcal{V}^d$ is the set of observed values in feature $d \in$ {W, -1W, +1W, P, -1P, +1P} and $\mathcal{V}^d$ is called feature d's feature values. Equation (7.6) is a FF that returns 1 when $\mathbf{x}_l$'s feature d equals to the feature value v^d and the variable y_l (lth label) equals to the label value i; otherwise 0.

We describe G and its feature function next, which also holds the key to the proposed L-CRF.

7.3.2   GENERAL DEPENDENCY FEATURE

The general dependency feature G uses generalized dependency relations. What is interesting about this feature is that it enables L-CRF to use past knowledge in its sequence prediction at the test time in order to perform better. This will become clear shortly. This feature takes a *dependency pattern* as its value, which is generalized from dependency relations.

The general dependency feature (G) of the variable $\mathbf{x}_l$ takes a set of feature values $\mathcal{V}^G$. Each feature value v^G is a dependency pattern. The *Label-G (LG)* FF is defined as:

$$f_{iv^G}^{LG}(y_l, \mathbf{x}_l) = \mathbb{1}\{y_l = i\}\mathbb{1}\{\mathbf{x}_l^G = v^G\}, \forall i \in \mathcal{Y}, \forall v^G \in \mathcal{V}^G \ . \tag{7.7}$$

Such a FF returns 1 when the dependency feature of the variable $\mathbf{x}_l$ equals to a dependency pattern v^G and the variable y_l equals to the label value i.

Dependency Relation

A dependency relation[2] is a quintuple-tuple: *(type, gov, govpos, dep, deppos)*, where *type* is the type of the dependency relation, *gov* is the *governor word*, *govpos* is the POS tag of the governor word, *dep* is the *dependent word*, and *deppos* is the POS tag of the dependent word. The l-th word can either be the governor word or the dependent word in a dependency relation.

Dependency Pattern

Dependency relations are generalized into *dependency patterns* using the following steps.

1. For each dependency relation, replace the current word (governor word or dependent word) and its POS tag with a wildcard since we already have the word (W) and the POS tag (P) features.

[2]Dependency relations are obtained using Stanford CoreNLP: http://stanfordnlp.github.io/CoreNLP/.

2. Replace the context word (the word other than the l-th word) in each dependency relation with a knowledge label to form a more general dependency pattern. Let the set of aspects annotated in the training data be K^t. If the context word in the dependency relation appears in K^t, we replace it with a knowledge label "A" (aspect); otherwise "O" (other).

For example, we work on the sentence "The battery of this camera is great." The dependency relations are given in Table 7.1. Assume the current word is "battery," and "camera" is annotated as an aspect. The original dependency relation between "camera" and "battery" produced by a parser is (nmod, battery, NN, camera, NN). Note that the word positions are not used in the relations in Table 7.1. Since the current word's information (the word itself and its POS-tag) in the dependency relation is redundant, it is replaced with a wild-card. The relation becomes (nmod, *, camera, NN). Secondly, since "camera" is in K^t, "camera" is replaced with a general label "A". The final dependency pattern becomes (nmod,*, A, NN).

Table 7.1: Dependency relations parsed from "The battery of this camera is great"

Index	Word	Dependency Relations
1	The	{(*det, battery, 2, NN, The, 1, DT*) }
2	battery	{(*nsubj, great, 7, JJ, battery, 2, NN*), (*det, battery, 2, NN, The, 1, DT*), (*nmod, battery, 2, NN, camera, 5, NN*) }
3	of	{(*case, camera, 5, NN, of, 3, IN*) }
4	this	{(*det, camera, 5, NN, this, 4, DT*) }
5	camera	{(*case, camera, 5, NN, of, 3, IN*), (*det, camera, 5, NN, this, 4, DT*), (*nmod, battery, 2, NN, camera, 5, NN*) }
6	is	{(*cop, great, 7, JJ, is, 6, VBZ*) }
7	great	{(*root, ROOT, 0, VBZ, great, 7, JJ*), (*nsubj, great, 7, JJ, battery, 2, NN*), (*cop, great, 7, JJ, is, 6, VBZ*) }

We now explain why dependency patterns can enable a CRF model to leverage the past knowledge. The key is the knowledge label "A" above, which indicates a likely aspect. Recall that our problem setting is that when we need to extract from the new domain D_{N+1} using a trained CRF model M, we have already extracted from many previous domains $D_1, \ldots, D_N$ and retained their extracted sets of aspects $A_1, \ldots, A_N$. Then, we can mine reliable aspects from $A_1, \ldots, A_N$ and add them in K^t, which adds many knowledge labels in the dependency patterns of the new data D_{N+1} due to sharing of aspects across domains. This enriches the dependency pattern features, which consequently allows more aspects to be extracted from the new domain data D_{N+1}.

7.3.3 THE L-CRF ALGORITHM

We now present the L-CRF algorithm. As the dependency patterns for the general dependency feature do not use any actual words and they can also use the prior knowledge, they are quite powerful for cross-domain extraction (the test domain is not used in training).

Let K be a set of *reliable aspects* mined from the aspects extracted in past domain datasets using the CRF model M. Note that it is assumed that M has already been trained using some labeled training data D^t. Initially, K is K^t (the set of all annotated aspects in the training data D^t). The more domains M has worked on, the more aspects it extracts, and the larger the set K gets. When faced with a new domain D_{N+1}, K allows the general dependency feature to generate more dependency patterns related to aspects due to more knowledge labels "A" as we explained in the previous section. Consequently, CRF has more informed features to produce better extraction results.

L-CRF works in two phases: *training phase* and *lifelong extraction phase*. The training phase trains a CRF model M using the training data D^t, which is the same as the normal CRF training, and will not be discussed further. In the lifelong extraction phase, M is used to extract aspects from incoming domains (M does not change and the domain data are unlabeled). All the results from the domains are retained in the past aspect store S. At a particular time, it is assumed M has been applied to N past domains, and is now faced with the $N + 1$ domain. L-CRF uses M and reliable aspects (denoted K_{N+1}) mined from S and K^t ($K = K^t \cup K_{N+1}$) to extract from D_{N+1}. Note that aspects K_t from the training data are considered always reliable as they are manually labeled, thus a subset of K. Not all extracted aspects from past domains can be used as reliable aspects due to many extraction errors. But those aspects that appear in multiple past domains are more likely to be correct. Thus, K contains those frequent aspects in S. The lifelong extraction phase is in Algorithm 7.13.

Lifelong Extraction Phase: Algorithm 7.13 performs extraction on D_{N+1} iteratively.

1. It generates features (F) on the data D_{N+1} (line 3), and applies the CRF model M on F to produce a set of aspects A_{N+1} (line 4).

2. A_{N+1} is added to S, the past aspect store. From S, we mine a set of frequent aspects K_{N+1}. The frequency threshold is λ.

3. If K_{N+1} is the same as K_p from the previous iteration, the algorithm exits as no new aspects can be found. We use an iterative process because each extraction gives new results, which may increase the size of K, the reliable past aspects or past knowledge. The increased K may produce more dependency patterns, which can enable more extractions.

4. Else: some additional reliable aspects are found. M may extract additional aspects in the next iteration. Lines 10 and 11 update the two sets for the next iteration.

The proposed L-CRF method was evaluated and compared with baselines. Experimental results show that L-CRF performs markedly better than CRF in the cross-domain setting, where one

---

Algorithm 7.13 Lifelong Extraction of L-CRF

1: $K_p \leftarrow \emptyset$
2: **loop**
3:     $F \leftarrow$ FeatureGeneration(D_{N+1}, K)
4:     $A_{N+1} \leftarrow$ Apply-CRF-Model(M, F)
5:     $S \leftarrow S \cup \{A_{N+1}\}$
6:     $K_{N+1} \leftarrow$ Frequent-Aspects-Mining(S, λ)
7:     **if** $K_p = K_{N+1}$ **then**
8:        **break**
9:     **else**
10:         $K \leftarrow K^t \cup K_{N+1}$
11:         $K_p \leftarrow K_{N+1}$
12:          $S \leftarrow S - \{A_{N+1}\}$
13:     **end if**
14: **end loop**

---

domain of data is used for training and the resulting model is tested on other domains. It also outperforms CRF in the in-domain setting, where the training and the testing data are from the same domain.

To close this section, we would like to point out that the technique presented in this section is highly limited because it does not update or improve the model itself. Moreover, learning on the job also has many types. For example, the problems solved in Chapter 8 and, to a great extent, in Chapter 5 can be regarded as learning on the job as well because they both involve learning while working on applications.

7.4   LIFELONG-RL: LIFELONG RELAXATION LABELING

In Shu et al. [2016], the authors proposed a *lifelong relaxation labeling* method called *Lifelong-RL* to incorporate LL in *relaxation labeling* [Hummel and Zucker, 1983] for belief propagation. Lifelong-RL was applied to a sentiment analysis task. This section gives an overview of Lifelong-RL. But first, we introduce the relaxation labeling algorithm. We then briefly describe how to incorporate the LL capability in relaxation labeling. For the application to the sentiment analysis task, please refer to the original paper.

7.4.1   RELAXATION LABELING

Relaxation Labeling is an unsupervised graph-based label propagation algorithm that works iteratively. The graph consists of nodes and edges. Each edge represents a binary relationship between two nodes. Each node n_i in the graph is associated with a multinomial distribution

$P(L(n_i))$ ($L(n_i)$ being the label of n_i) on a label set Y. Each edge is associated with two conditional probability distributions $P(L(n_i)|L(n_j))$ and $P(L(n_j)|L(n_i))$, where $P(L(n_i)|L(n_j))$ represents how the label $L(n_j)$ influences the label $L(n_i)$ and vice versa. The neighbors $Ne(n_i)$ of a node n_i are associated with a weight distribution $w(n_j|n_i)$ with $\sum_{n_j \in Ne(n_i)} w(n_j|n_i) = 1$.

Given the initial values of these quantities as inputs, relaxation labeling iteratively updates the label distribution of each node until convergence. Initially, we have $P^0(L(n_i))$. Let $\Delta P^{r+1}(L(n_i))$ be the change of $P(L(n_i))$ at iteration $r + 1$. Given $P^r(L(n_i))$ at iteration r, $\Delta P^{r+1}(L(n_i))$ is computed by:

$$\Delta P^{r+1}(L(n_i)) = \sum_{n_j \in Ne(n_i)} \left(w(n_j|n_i) \times \sum_{y \in Y} P(L(n_i)|L(n_j) = y) \times P^r(L(n_j) = y) \right). \quad (7.8)$$

Then, the updated label distribution for iteration $r + 1$, $P^{r+1}(L(n_i))$, is computed with:

$$P^{r+1}(L(n_i)) = \frac{P^r(L(n_i)) \times (1 + \Delta P^{r+1}(L(n_i)))}{\sum_{y \in Y} P^r(L(n_i) = y) \times (1 + \Delta P^{r+1}(L(n_i) = y))}. \quad (7.9)$$

Once relaxation labeling ends, the final label of node n_i is its highest probable label: $L(n_i) = \underset{y \in Y}{\mathrm{argmax}}(P(L(n_i) = y))$.

Note that $P(L(n_i)|L(n_j))$ and $w(n_j|n_i)$ are not updated in each relaxation labeling iteration but only $P(L(n_i))$ is. $P(L(n_i)|L(n_j))$, $w(n_j|n_i)$, and $P^0(L(n_i))$ are provided by the user or computed based on the application context. Relaxation labeling uses these values as input and iteratively updates $P(L(n_i))$ based on Equations (7.8) and (7.9) until convergence. Next we discuss how to incorporate LL in relaxation labeling.

7.4.2   LIFELONG RELAXATION LABELING

For LL, as usual, it is assumed that at any time step, the system has worked on N past domain data $\mathcal{D}^p = \{\mathcal{D}_1, \mathcal{D}_2, \ldots, \mathcal{D}_N\}$. For each past domain data $\mathcal{D}_i \in \mathcal{D}^N$, the same Lifelong-RL algorithm has been applied and its result has been saved in the knowledge base (KB). Then the algorithm can borrow some useful prior/past knowledge in the KB to help relaxation labeling in the new/current domain $\mathcal{D}_{N+1}$. Once the result of the current domain is produced, it is also added to the KB for future use.

We now discuss the specific types of information or knowledge that can be obtained from the previous tasks to help relaxation labeling in the future, which are thus stored in the KB.

1. *Prior edges*: In many applications, the graph is not given. Instead, it has to be constructed based on the data from the new task/domain data $\mathcal{D}_{N+1}$. However, due to the limited data in $\mathcal{D}_{N+1}$, some edges between nodes that should be present are not extracted from the data. But such edges between the nodes may exist in some past domain data. Then, those edges and their associated probabilities can be borrowed.

2. *Prior labels*: Some nodes in the new task/domain may also exist in some previous tasks/domains. Their labels in the past domains are very likely to be the same as those in the current domain. Then, those prior labels can give us a better idea about the initial label probability distributions of the nodes in the current domain $\mathcal{D}_{N+1}$.

To leverage those edges and labels from the past domains, the system needs to ensure that they are likely to be correct and applicable to the current task. This is a challenging problem. Interested readers, please refer to the original paper [Shu et al., 2016].

7.5 SUMMARY AND EVALUATION DATASETS

Information extraction is naturally a continuous process as there is always knowledge to be extracted. It is also natural that the knowledge extracted earlier can help subsequent extractions. In this chapter, we first described the well-known NELL system as an excellent example of life-long semi-supervised information extraction system. We introduced its key ideas, architecture, and various sub-systems and algorithms. It is by no means exhaustive as many specific aspects of the system were not given in-depth treatments. The system is also continuously evolving and is becoming more and more powerful. What is very valuable about NELL is that it is perhaps the only non-stop or continuous learning system that extracts information from both unstructured text and semi-structure documents on the Web. We believe that more such real-life LL systems should be constructed to truly realize continuous learning, knowledge accumulation, and problem solving. Such systems will allow researchers to gain true insights into LL on how LL may work in practice and what the technical challenges are. These insights will help us design better and more practically useful LL systems and techniques.

This chapter also discussed three other papers. One is about opinion target (or aspect) extraction using LL. Its LL idea is based on multi-domain recommendation, which is essentially a meta-mining method. One is about learning after model building. This is a new idea because in traditional learning, after a model is built, it is simply applied to applications and there is no further learning during the application process. Finally, the chapter discussed a belief propagation method with the help of LL. Here, we saw that the past knowledge can provide more accurate prior probabilities and also help expand the graph itself to provide more information for more accurate propagation.

Regarding experimental datasets, NELL uses webpages continuously crawled from the Web. The other papers used product reviews. The product review datasets have been listed in the last section of Chapter 6. Two datasets have been used in the evaluation of the techniques proposed in Liu et al. [2016] and Shu et al. [2017b]. These are aspect-annotated datasets and are publicly available. One has five review collections and the other has three review collections [Liu et al., 2015a].[3]

---

[3]https://www.cs.uic.edu/~liub/FBS/sentiment-analysis.htm

C H A P T E R 8

Continuous Knowledge Learning in Chatbots

This chapter discusses the emerging research topic of *lifelong interactive knowledge learning* for chatbots [Mazumder et al., 2018]. Continuous learning in an interactive environment is a key capability of human beings. One can only learn so much by being told or supervised because the world is simply too complex to be completely learned this way. In fact, we humans probably learn a great deal of our knowledge through interactions with other humans and the environment around us which constantly give us explicit and implicit feedback. This learning process is called *self-supervised* because it does not require human annotated/labeled training data. In the context of chatbots, lifelong interactive learning is critical because in order for a chatboot to be truly intelligent in human-machine conversation, it has to continually learn new knowledge in order to improve itself and to understand and to get to know each of its conversation partners. Note that we use the term chatbots to refer to all kinds of conversational agents, such as dialogue systems and question-answering systems.

Chatbots have a long history in AI and natural language processing (NLP). They became particularly and increasingly popular in the past few years due to the commercial success of some chatbots or virtual assistants such as Echo and Siri. Numerous chatbots have been developed or are under development, and many researchers are also actively working on techniques for chatbots.

Early chatbot systems were mostly built using markup languages such as AIML,[1] hand-crafted conversation generation rules, and/or information retrieval techniques [Ameixa et al., 2014, Banchs and Li, 2012, Lowe et al., 2015, Serban et al., 2015]. Recent neural conversation models [Li et al., 2017b, Vinyals and Le, 2015, Xing et al., 2017] are able to perform some limited open-ended conversations. However, since they do not use explicit knowledge bases (KBs) and do not perform inference, they often suffer from generic and dull responses [Li et al., 2017a, Xing et al., 2017]. More recently, Ghazvininejad et al. [2017] and Le et al. [2016] proposed to use KBs to help generate responses for knowledge-grounded conversation. However, a major weakness of the existing chat systems is that they do learn new knowledge during conversation, i.e., their knowledge is fixed beforehand and cannot be expanded or updated during the conversation process. This seriously limits the scope of their applications. Even though some existing systems can use very large KBs, these KBs still miss a large number of facts (knowledge) [West

---

[1]http://www.alicebot.org/

et al., 2014]. It is thus important for a chatbot to continuously learn new knowledge in the conversation process to expand its KB and to improve its conversation capability, i.e., learning on the job.

Since there is little work in this emerging area, this chapter presents only one work that aims to build a lifelong interactive knowledge learning engine for chatbots [Mazumder et al., 2018]. The goal of the engine is to learn a specific type of knowledge called *factual knowledge* during the interactive conversation process. A piece of such knowledge is called a *fact* and is represented as a triple: (s, r, t), which says that the source entity s and target entity t are linked by the relation r. For example, (*Obama*, *CitizenOf*, *USA*) means that Obama is a citizen of USA.

8.1 LILI: LIFELONG INTERACTIVE LEARNING AND INFERENCE

Mazumder et al. [2018] modeled the interactive knowledge learning as an *open-world knowledge base completion* problem, which is an extension to the traditional *knowledge base completion* (KBC) problem. KBC aims to infer new facts (knowledge) automatically from the existing facts in a given KB. It is defined as a binary classification problem. Given a query triple, (s, r, t), we predict whether the source entity s and target entity t can be linked by the relation r. Previous approaches [Bordes et al., 2011, 2013, Lao et al., 2011, 2015, Mazumder and Liu, 2017, Nickel et al., 2015] solve this problem under the *closed-world* assumption, i.e., s, r and t are all *known* to exist in the KB. This is a major weakness because it means that no new knowledge or facts may contain unknown entities or relations. Due to this limitation, KBC is not sufficient for knowledge learning in conversations because in a conversation, the user can say anything, which may contain entities and/or relations that are not already in the existing KB.

Mazumder et al. [2018] removed the closed-world assumption of KBC, and allowed all s, r, and t to be *unknown*. The new problem is called *open-world knowledge base completion* (OKBC). OKBC generalizes KBC and can naturally serves as a model for knowledge learning in conversation. In essence, OKBC is an *abstraction* of the knowledge learning and inference problem in conversation. The problem is solved in the open and interactive conversation process.

The paper claimed that in a conversation, two key types of factual information, true facts and queries, can be extracted from the user utterance. Queries are facts whose truth values need to be determined. The work does not deal with subjective information such as beliefs and opinions. The paper also does not study fact or relation extraction from natural language text (user utterances) as there has been an extensive work on these topics in natural language processing (NLP). It assumes that an extraction system is already in place.

Mazumder et al. [2018] dealt with the two types of information as follows. For a true fact, it is incorporated into the KB. Here the system needs to make sure that it is not already in the KB, which involves relation resolution and entity linking. The paper again assumes that an existing system can be used to do this. After a fact is added to the KB, it may predict that some related facts involving some existing relations in the KB may also be true. For example, if the user says "Obama was born in USA," the system may guess that (*Obama*, *CitizenOf*, *USA*) (meaning that Obama is a citizen of USA) could also be true based on the current KB. To verify this fact, it needs to solve a KBC problem by treating (*Obama*, *CitizenOf*, *USA*) as a query. This is a KBC problem because the fact (*Obama*, *BornIn*, *USA*) extracted from the original sentence has been added to the KB. Then Obama and U.S. are in the KB. If the KBC problem is solved, it learns a new fact (Obama, CitizenOf, USA) in addition to the extracted fact (Obama, BornIn, USA). For a query fact, e.g., (Obama, BornIn, USA), extracted from a user question "Was Obama born in USA?" the system needs to solve an OKBC problem if any of "*Obama*," "*BornIn*," or "*USA*" is not already in the KB.

We can see that OKBC is the core problem of a knowledge learning engine for conversation. Thus, Mazumder et al. [2018] focused on solving the OKBC problem. It assumes that other tasks such as fact/relation extraction and resolution and inferring related facts of an extracted fact are solved by other sub-systems or existing algorithms.

The paper solves the OKBC problem by mimicking how humans acquire knowledge and perform reasoning in an interactive conversation. Whenever we humans encounter an unknown concept or relation while answering a query, we perform inference using our existing knowledge. If our knowledge does not allow us to draw a conclusion, we typically ask questions to others to acquire the related knowledge and use it in the inference. The process typically involves an *inference strategy* (a sequence of actions), which interleaves a sequence of *processing* and *interactive* actions. A processing action can be the selection of related facts, deriving inference chain, etc., that advances the inference process. An interactive action can be deciding what to ask, formulating a suitable question to ask, etc., that enable us to interact. The process helps grow the knowledge over time and the newly gained knowledge enables the system to communicate better in the future. This process is called *lifelong interactive learning and inference* (LiLi). Lifelong learning is reflected by the fact that the newly acquired facts are retained in the KB and used in inference for future queries, and that the accumulated knowledge in addition to the updated KB including past inference performances are leveraged to guide future interactions and learning. LiLi has the following capabilities:

1. ***formulating an inference strategy*** for a given query that embeds *processing* and *interactive* actions;

2. ***learning interaction behaviors*** (deciding what to ask and when to ask the user);

3. ***leveraging the acquired knowledge*** in the current and future inference process; and

4. ***performing 1, 2, and 3 in a lifelong manner*** for continuous knowledge learning.

LiLi starts with the closed-world KBC approach *path-ranking* (PR) [Gardner and Mitchell, 2015, Lao et al., 2011] and extends KBC to open-world knowledge base completion (OKBC). For a relation r, PR works by enumerating paths (except single-link path r) between entity-pairs linked by r in the KB and use them as features to train a binary classifier to predict whether a query (s, r, t) should be in the KB. Here, a path between two entities is a sequence of relations linking them. In Mazumder et al. [2018], the latest PR method, C-PR [Mazumder and Liu, 2017], was adopted and extended to make it work in the open-world setting. C-PR enumerates paths by performing bidirectional random walks over the KB graph while leveraging the context of the source-target entity-pair. It also adopts and extends the compositional vector space model in Neelakantan et al. [2015] and Das et al. [2016] with continuous learning capability for prediction.

User:	(Obama, CitizenOf, USA)? "Is Obama a citizen of USA?" [Query]
LiLi:	I do not know what "**CitizenOf**" mean? Can you provide me an example? [*Ask for Clue*]
User:	(David Cameron, CitizenOf, UK). "David Cameron is a citizen of UK." [SF1]
LiLi:	Got it. Can you tell me how "**Obama**" and "**Honolulu**" are related? [*CLQ*]
User:	(Obama, BornIn, Honolulu). "Obama was born in Honolulu." [SF2]
LiLi:	Got it. Can you tell also me how "**Honolulu**" and "**Hawaii**" are related? [*MLQ*]
User:	(Honolulu, CapitalOfState, Hawaii). "Honolulu is the state capital of Hawaii." [SF3]
LiLi:	(Obama, CitizenOf, USA) is correct. [Answer]

Figure 8.1: An example of interactive inference and learning. Note that LiLi only works with triples. Each triple above is assumed to be extracted from the sentence after it.

Given an OKBC query (s, r, t) (e.g., "(*Obama, CitizenOf, USA*)," which means whether Obama a citizen of USA), LiLi interacts with the user (*if needed*) by *dynamically* formulating questions (see the interaction example in Figure 8.1[2]) and leverages the interactively acquired knowledge (supporting facts (SFs) in the figure) for continued inference. To do so, LiLi formulates a query-specific inference strategy and executes it. LiLi is designed in a Reinforcement Learning (RL) setting that performs sub-tasks like formulating and executing a strategy, training a prediction model for inference, and retaining the knowledge for future use. The effectiveness of LiLi was empirically verified using two *standard* real-world KBs: *Freebase*[3] and *WordNet*.[3]

8.2   BASIC IDEAS OF LILI

As explained above, OKBC naturally serves as a model for knowledge learning in conversation. The question now is how to solve the OKBC problem. The key idea in Mazumder et al. [2018] is to map OKBC to KBC through interaction with the user by asking user questions. KBC already has existing solutions, e.g., C-PR.

[2]Note that the user query and system response are represented as triples as this paper does not build a conversation system. It only builds a core knowledge acquisition engine. Also, the query may be from a user or a system, e.g., a question-answer system or a conversation system that has extracted a candidate fact and wants to verify it and add it to the KB. This paper also does not study the case that the query fact is already in the KB, which is easy to verify. Moreover, as this work focuses on knowledge learning and inference *rather than conversation modeling* it simply uses template-based question generation to model LiLi's interaction with the user.

[3]https://everest.hds.utc.fr/doku.php?id=en:smemlj12

Mapping open-world to close-world. Clearly, the closed-world model KBC cannot solve the open-world OKBC problem. For example, the KBC method C-PR cannot extract path features and learn a prediction model when any of s, r, or t is unknown. LiLi solves this problem through *interactive* knowledge acquisition. If r is *unknown*, LiLi asks the user to provide a clue (an example of r). And if s or t is *unknown*, LiLi asks the user to provide a link (relation) to connect the unknown entity with an existing entity (automatically selected) in the KB. Such a query is referred to as a *connecting link query* (CLQ). The acquired knowledge, which basically makes s, r, and t known in the KB, reduces OKBC to KBC and makes the C-PR inference task feasible.

LiLi is designed as a combination of two interconnected models.

1. An RL model that learns to formulate a query-specific inference strategy for performing the OKBC task. LiLi's strategy formulation is modeled as a Markov Decision Process (MDP) with finite state ($\mathcal{S}$) and action ($\mathcal{A}$) spaces. A state $S \in \mathcal{S}$ consists of 10 binary state variables (Table 8.1), each of which keeps track of the results of an action $a \in \mathcal{A}$ (Table 8.2) taken by LiLi and thus, records the progress made in the inference process so far. We can see actions for user interactions in Table 8.2 that can turn the OKBC problem into a KBC problem. The RL algorithm Q-learning [Watkins and Dayan, 1992] with ϵ-greedy strategy is used to learn the optimal policy for training the RL model.

2. A lifelong prediction model for predicting whether a triple should be in the KB, which is invoked by an action while executing the inference strategy, is learned for each relation as in C-PR. LiLi uses deep learning to build the model. Since a model trained on a few examples (e.g., clues acquired for unknown r) with randomly initialized weights of the neural network model often perform poorly due to underfitting, it transfers the knowledge (weights) from the past most similar (with regard to r) task in an LL manner. LiLi uses a relation-entity matrix $\mathcal{M}$ to find the past most similar task for r (discussed below). See Mazumder et al. [2018] for more details.

The framework improves its performance over time through user interaction and knowledge retention. Compared to the existing KB inference methods, LiLi overcomes the following two challenges for OKBC.

1. **Spareseness of KB.** A main issue of all PR methods like C-PR is the connectivity of the KB graph. If there is no path connecting s and t in the graph, path enumeration of C-PR gets stuck and inference becomes infeasible. In such cases, LiLi uses a template relation ("@-?-@") as the *missing link* marker to connect entity-pairs and continues feature extraction. A path containing "@-?-@" is called an *incomplete path*. Thus, the extracted feature set contains both complete (no missing link) and incomplete paths. Next, LiLi selects an incomplete path from the feature set and asks the user to provide a link for path completion. Such a query is referred to as *missing link query* (MLQ).

2. **Limitation in user knowledge.** If the user is unable to respond to MLQs or CLQs, LiLi uses a *guessing mechanism* to fill the gap. This enables LiLi to continue its inference even if the user cannot answer a system's question.

Table 8.1: State bits and their meanings.

State Bits	Name	Description
QERS	Query Entities and Relation Searched	Whether the query source (s) and target (t) entities and query relation (r) have been searched in KB or not
SEF	Source Entity Found	Whether the source entity (s) has been found in KB or not
TEF	Target Entity Found	Whether the target entity (t) has been found in KB or not
QRF	Query Relation Found	Whether the query relation (r) has been found in KB or not
CLUE	Clue Bit Set	Whether the query is a clue or not
ILO	Interaction Limit Over	Whether the interaction limit is over for the query or not
PFE	Path Feature Extracted	Whether path feature extraction has been done or not
NEFS	Non-empty Feature Set	Whether the extracted feature set is non-empty or empty
CPF	Complete Path Found	Whether the extracted path features are complete or not
INFI	Inference Invoked	Whether Inference instruction has been invoked or not

Table 8.2: Actions and their descriptions.

ID	Description
a_0	Search source (h), target (t) entities and query relation (r) in KB
a_1	Ask user to provide an example/clue for query relation r
a_2	Ask user to provide missing link for path feature completion
a_3	Ask user to provide the connecting link for augmenting a new entity with KB
a_4	Extract path features between source (s) and target (t) entities using C-PR
a_5	Store query data instance in data buffer and invoke prediction model for inference

8.3 COMPONENTS OF LILI

As LL needs to retain knowledge learned from past tasks and use it to help future learning, LiLi uses a *Knowledge Store (KS)* for knowledge retention. KS has four components:

1. **Knowledge Graph** (G): G (the KB) is initialized with base KB triples and gets expanded and updated over time with the acquired knowledge.

2. **Relation-Entity Matrix** ($\mathcal{M}$): $\mathcal{M}$ is a sparse matrix, with rows as relations and columns as entity-pairs and is used by the prediction model. Given a triple $(s, r, t) \in G$, we set $\mathcal{M}[r, (s, t)] = 1$ indicating r occurs for pair (s, t).

3. **Task Experience Store** ($\mathcal{T}$): $\mathcal{T}$ stores the predictive performance of LiLi on past learned tasks in terms of *Matthews correlation coefficient* (MCC)[4] that measures the quality of binary classification. So, for two tasks r and r' (each relation indicates a task), if $\mathcal{T}[r] > \mathcal{T}[r']$ [where $\mathcal{T}[r]$=MCC(r)], we say C-PR has learned r well compared to r'.

4. **Incomplete Feature DB** (Π_{DB}): Π_{DB} stores the frequency of an incomplete path π in the form of a tuple (r, π, e_{ij}^{π}) and is used in formulating MLQs. $\Pi_{DB}[(r, \pi, e_{ij}^{\pi})] = N$ implies LiLi has extracted incomplete path π N times involving entity-pair e_{ij}^{π} [(e_i, e_j)] for query relation r.

The RL model learns even after training whenever it encounters an unseen state (in testing) and thus, gets updated over time. KS is updated continuously over time as a result of the execution of LiLi and takes part in future learning. The prediction model uses LL, where we transfer knowledge (parameter values) from the model for the past most similar task to help learn the current task. Similar tasks are identified by factorizing $\mathcal{M}$ and computing a task similarity matrix $\mathcal{M}_{sim}$. Besides LL, LiLi uses $\mathcal{T}$ to identify poorly learned past tasks and acquire more clues for them to improve its skillset over time.

LiLi also uses a stack, called *Inference Stack* ($\mathcal{IS}$) to hold query and its state information for RL. LiLi always processes stack top ($\mathcal{IS}[top]$). The clues from the user get stored in $\mathcal{IS}$ on top of the query during strategy execution and processed first. Thus, the prediction model for r is learned before performing inference on a query, transforming OKBC to a KBC problem.

8.4   A RUNNING EXAMPLE

The working of LiLi is involved. For details, please refer to Mazumder et al. [2018]. Here we provide a running example by working on the example shown in Figure 8.1. LiLi works on the example as follows: First, LiLi executes a_0 and detects that the source entity "*Obama*" and query relation "*CitizenOf*" are *unknown*. Thus, LiLi executes a_1 to acquire clue (SF1) for "*CitizenOf*" and pushes the clue (+ve example) and two generated -ve examples into $\mathcal{IS}$. Once the clues are processed and a prediction model is trained for "*CitizenOf*" by formulating separate strategies for them, LiLi becomes aware of "*CitizenOf*." Now, as the clues have already been popped from $\mathcal{IS}$, the query becomes $\mathcal{IS}[top]$ and the strategy formulation process for the query resumes. Next, LiLi asks user to provide a connecting link for "*Obama*" by performing a_3. Now, the query entities and relation being known, LiLi enumerates paths between "*Obama*" and "*USA*" by performing a_4. Let an extracted path be "$Obama - BornIn \rightarrow Honolulu - @-? - @ \rightarrow Hawaii - StateOf \rightarrow USA$" with missing link between ($Honolulu, Hawaii$). LiLi asks

---
[4]https://en.wikipedia.org/wiki/Matthews_correlation_coefficient

the user to fill the link by performing a_2 and then, extracts the complete feature "$BornIn \rightarrow CapitalOfState \rightarrow StateOf$." The feature set is then fed to the prediction model and inference is made as a result of a_5. Thus, the formulated inference strategy is: "$\langle a_0, a_1, a_3, a_4, a_2, a_5 \rangle$."

8.5   SUMMARY AND EVALUATION DATASETS

In this chapter, we discussed an initial attempt to build an engine for continuous knowledge learning in human-machine conversation. We first showed that the problem underlying the engine can be formulated as an open-world knowledge base completion (OKBC) problem. We then briefly described the lifelong interactive learning and inference (LiLi) approach to solving the OKBC problem. OKBC is a generalization of KBC (knowledge base completion). LiLi solves the OKBC problem by mapping OKBC to KBC through interacting with the user. The process is formulated as a query-specific inference strategy and modeled as and learned through RL. The resulting strategy is then executed to solve the problem which involves interacting with the user in a lifelong manner.

This work, however, is still preliminary, and has several weaknesses. First, it is not integrated with a chatbot system, and it assumes that the tasks of relation extraction, resolution, entity linking, etc., can be done by existing techniques. However, although there are many existing techniques for them, these tasks are still very challenging. Second, it is only designed for learning factual knowledge that can be expressed as triples. Many other forms of knowledge are not considered.

About evaluation datasets, three well-known KBs (1) FB15k,[5] (2) WordNet,[8] and (3) ConceptNet[6] are used in Mazumder et al. [2018]. For candidate fact extraction and conversation generation, one can learn using (1) a traditional relation extraction dataset such as the one used in TAC KBP Slot Filling challenge [Angeli et al., 2015], and (2) publicly available benchmark conversation datasets such as the Ubuntu dialogue corpus [Lowe et al., 2015], Cornell Movie—Dialogs Corpus[7] Danescu-Niculescu-Mizil and Lee [2011] and Wikipedia Talk Page Conversations Corpus[8] Danescu-Niculescu-Mizil et al. [2012], respectively.

---

[5]https://everest.hds.utc.fr/doku.php?id=en:smemlj12
[6]https://github.com/commonsense/conceptnet5/wiki/Downloads
[7]http://www.cs.cornell.edu/~cristian/Cornell_Movie-Dialogs_Corpus.html
[8]http://www.cs.cornell.edu/~cristian/Echoes_of_power.html

CHAPTER 9

Lifelong Reinforcement Learning

This chapter discusses *lifelong reinforcement learning*. Reinforcement learning (RL) is the problem where an agent learns actions through trial-and-error interactions with a dynamic environment [Kaelbling et al., 1996, Sutton and Barto, 1998]. In each interaction step, the agent receives input on the current state of the environment. It chooses an action from a set of possible actions. The action changes the state of the environment. Then, the agent gets the value of this state transition, which can be a reward or penalty. This process repeats as the agent learns a trajectory of actions to optimize its objective, e.g., to maximize the long-run sum of rewards. The goal of RL is to learn an *optimal policy* that maps states to actions (possibly stochastically). There is a recent surge in research in RL due to its successful use in the computer program called *AlphaGo* [Silver et al., 2016], which won 4–1 against one of the legendary professional Go players Lee Sedol in March 2016.[1] More recently, AlphaGo Zero [Silver et al., 2017][2] was designed to learn to master the game of Go from scratch without human knowledge, and it has achieved superhuman performance.

Let us see an example of a RL setting [Tanaka and Yamamura, 1997]. This example involves an agent trying to find gold in an $N \times N$ gridworld maze. The agent can choose one action from a set of possible actions, moving left/right/up/down and picking up an item. The maze, which is the environment, may have obstacles, monsters, and gold. When the agent picks up the gold, it gets a positive reward (say +1,000). If the agent is killed by a monster, it gets a negative reward (say −1,000). When the agent steps into an obstacle, it will retreat to the previous location. The agent keeps interacting with the environment through actions and reward feedback to learn the best sequence of actions. The goal is to maximize the total reward (final reward—cost of all actions taken).

RL is different from supervised learning in that there is no input/output pair in RL. In supervised learning, the manual label indicates the best output label for an input. However, in RL, after an action is taken, the agent is *not* told which action would have been in its best long-term interests. So the agent needs to gain useful experience and learn an optimal sequence of actions through interactions with the environment via feedback.

---

[1]https://deepmind.com/alpha-go
[2]https://deepmind.com/blog/alphago-zero-learning-scratch/

However, in order to achieve high-quality performance, the agent usually needs a large amount of quality experience. This is particularly true in high-dimensional control problems. The high cost of gaining such experience is a challenging issue. In order to overcome it, *lifelong reinforcement learning* (*lifelong RL*) has been proposed and studied by several researchers. The motivation is to use the experience accumulated from other tasks to improve the agent's decision making in the current new task.

Lifelong RL was first proposed by Thrun and Mitchell [1995] who worked on a lifelong robot learning problem. They showed that with knowledge memorization, the robot can learn faster while relying less on real-world experimentations. Ring [1998] proposed a continual-learning agent that aims to gradually solve complicated tasks by learning easy tasks first. Tanaka and Yamamura [1997] treated each environment as a task, and constructed an artificial neural network for each task/environment. They then used the weights of the nodes in the neural networks for existing tasks to initialize the neural network for the new task. Konidaris and Barto [2006] proposed to use approximations of prior optimal value functions for initialization in a new task. The intuition is that an agent can be trained on a sequence of relatively easy tasks to gain experience and develop a more informative measure of reward, which can then be leveraged when performing harder tasks. Wilson et al. [2007] proposed a hierarchical Bayesian lifelong RL technique in the framework of Markov Decision Process (MDP). In particular, they added a random variable to indicate MDP classes, and assumed that the MDP tasks being assigned to the same class are similar to each other. A nonparametric infinite mixture model was proposed to take into account the unknown number of MDP classes. Fernández and Veloso [2013] proposed a policy reuse method in lifelong RL where policies learned from prior tasks are probabilistically reused to help a new task. A nonlinear feedback policy that generalizes across multiple tasks is also used as knowledge in Deisenroth et al. [2014]. Knowledge policy is defined as a function of both state and task, which can account for unknown states in an existing task and states in a new task. Brunskill and Li [2014] studied lifelong RL via PAC-inspired option discovery. They showed that the learned options from previous experience can potentially accelerate learning in the new task. Bou Ammar et al. [2014] proposed a Policy Gradient Efficient Lifelong Learning Algorithm (PG-ELLA) that extends ELLA [Ruvolo and Eaton, 2013b] for lifelong RL. Along the same line, Bou Ammar et al. [2015a] proposed a cross-domain lifelong reinforcement learner based on policy gradient methods. Later, Bou Ammar et al. [2015c] added constraints to PG-ELLA for safe lifelong learning. Tessler et al. [2017] proposed a lifelong learning system that transfers reusable skills to solve tasks in Minecraft (a video game). The knowledge is represented by deep networks. Tutunov et al. [2017] proposed a distributed Newton method for lifelong policy search. Zhan et al. [2017] focused on scalability of lifelong RL and proposed an algorithm to reach linear convergence rates in operations. El Bsat and Taylor [2017] proposed a multitask policy search framework that also achieves linear convergence speed. This chapter reviews the representative techniques proposed for lifelong RL.

9.1 LIFELONG REINFORCEMENT LEARNING THROUGH MULTIPLE ENVIRONMENTS

Tanaka and Yamamura [1997] proposed a lifelong RL technique that treats an environment as a task. In their problem setting, there is a set of tasks, i.e., a set of environments. The tasks are independent of one anther. For example, there is a set of mazes and each maze setting is an environment. In each of the mazes, the places of start and gold are fixed while other environment factors such as the places of obstacles and monsters, or the maze size, are different. Clearly, the environments and the tasks are assumed to share some common properties.

A two-step algorithm was proposed for learning [Tanaka and Yamamura, 1997]: (a) acquiring bias from previous N tasks and (b) incorporating bias into the new $(N + 1)$th task. The bias here is the knowledge in the LL context to be exploited. The bias consists of two parts: initial bias and learning bias. The initial bias is used to initialize the model starting stage. The learning bias is used to influence the modeling or learning process. A neural network was used as an example model in this work. To incorporate bias, the authors applied a stochastic gradient method [Kimura, 1995] with a new update equation. The details are discussed in the subsection below.

9.1.1 ACQUIRING AND INCORPORATING BIAS

For each task/environment, a neural network is constructed. To simplify the model, the authors used a two-layer neural network. In each task t, each neural network node (i, j) has a weight $w_{i,j}^t$. The intuition is that if the weight of a node does not change much throughout the learning process of the tasks, it can be used as an invariant node. On the other hand, if the weight of a node varies a lot, it is likely to be a task-dependent node.

Based on this idea, two types of biases are acquired from the previous N tasks and they are then applied in the learning phrase of the new $(N + 1)$th task.

1. *Initial bias*: In RL, the initial random walk stage is usually very expensive. It is thus important to have a good initialization in order to improve the speed of convergence and the final performance. Initial bias is used to provide a good initial stage in order to reduce this cost. The initial weight of a node (i, j) for the $(N + 1)$th task is the average weight of the same node across all the previous N tasks, i.e., $\frac{1}{N} \sum_{t=1}^{N} w_{i,j}^t$.

2. *Learning bias*: Since the stochastic gradient method [Kimura, 1995] is used, the weight of each node can have a different learning rate based on their variance in the previous tasks. Following this intuition, those nodes that have varying weights in the previous tasks are more likely to be task dependent, and thus require slightly larger learning rates than those nodes with little weight changes. So for a node (i, j) in the $(N + 1)$th task, its weight update is performed as follows:

$$w_{i,j}^{N+1} \leftarrow w_{i,j}^{N+1} + \alpha\beta_{i,j}(1-\gamma)\Delta w_{i,j}^{N+1}, \text{ and} \qquad (9.1)$$

$$\beta_{i,j} = \epsilon \left(1 + \max_{t=1,\ldots,N} w_{i,j}^{N+1} - \min_{t=1,\ldots,N} w_{i,j}^{N+1} \right) \ . \tag{9.2}$$

Here α is the universal learning rate for all nodes. $\beta_{i,j}$ is the learning bias for each node and it controls the learning rate. ϵ is the bias parameter.

In a nutshell, the neural network for the $(N + 1)$th task is initialized with *initial bias* and then updated via *learning bias* with gradient-updating equations of Equations (9.1) and (9.2).

9.2   HIERARCHICAL BAYESIAN LIFELONG REINFORCEMENT LEARNING

Wilson et al. [2007] worked on RL in the framework of Markov Decision Process (MDP). The way to solve an MDP problem is to find an optimal policy that minimizes the total expected costs/penalties. Instead of working on only one MDP task in isolation, the authors considered a sequence of MDP tasks, and proposed a model called MTRL (Multi-Task Reinforcement Learning). Although the term *multi-task* is used in the name, MTRL is in fact an online multi-task learning (MTL) method, which is considered as an LL method. The key idea of MTRL is the use of the hierarchical Bayesian approach to model "classes" of MDPs. Each class (or cluster) has some shared structure, which is regarded as the shared knowledge and is transferred to a new MDP of the class. This strong prior makes the learning of the new MDP much more efficient.

9.2.1   MOTIVATION

This work assumes that the MDP tasks are chosen randomly from a fixed but unknown distribution [Wilson et al., 2007]. As a result, the MDP tasks share some aspects that enable the knowledge extraction and transfer. To understand why the shared aspects may help the agent more quickly learn the optimal policy for a new MDP task, let's follow the gold-finding example at the beginning of the chapter.

Each MDP task is to find gold in a maze. The maze may contain obstacles, monsters, and gold. Depending on the type of environment, certain types of rocks might be good indicators of the presence of gold while some other types of rocks may be correlated with the absence of gold. Also, some signals such as noise or smell may come from monsters nearby. If an agent learns everything from scratch, it may take a long time to learn all these rules and adjust its behaviors. However, with the observations from previous MDP tasks, the agent may learn some useful knowledge, e.g., some monsters carry a strong smell. Using such knowledge, the agent can quickly adjust itself to avoid this type of monster when it detects the smell. The idea is that given the knowledge from the previous MDP tasks and a small amount of experience in the new MDP task, the agent can exploit the knowledge to explore the new MDP environment much more efficiently.

9.2.2   HIERARCHICAL BAYESIAN APPROACH

Bayesian modeling was applied to tackle the problem in the paper. In the single-task scenario, a Bayesian model-based RL computes the posterior distribution $P(M|\Theta, \mathcal{O})$ where M denotes a random variable over MDPs. $\mathcal{O}$ is the observation set and Θ is the set of model parameters. This distribution is used to help the agent choose actions. It will evolve with more actions and observations. One naïve way to extend this single-task approach to LL is to assume that all the MDP tasks are the same and treat the observations as coming from a single MDP task. Obviously, if the MDP tasks are not the same, this naïve method does not perform well.

To consider the differences between MDP tasks, Wilson et al. [2007] proposed a hierarchical Bayesian model that adds a random variable C to indicate MDP classes (or groups of similar MDPs). The assumption is that the MDP tasks within the same class assignment are similar to each other while the MDP tasks with different class assignments are very different from each other. Here, $\mathcal{M}$ denotes an MDP task and M denotes a random variable over MDPs. The sequence of MDP tasks are represented by $\mathcal{M}_1, \mathcal{M}_2, \ldots$. Instead of having the posterior distribution as $P(M|\Theta, \mathcal{O})$ in the single-task case, the posterior distribution for ith task in the hierarchical case is modeled as $P(M|\Psi, \mathcal{O}_i)$ where $\Psi = \{\Theta, \mathcal{C}\}$. Θ denotes the parameters under each class and $\mathcal{C}$ means all class assignments. $\mathcal{O}_i$ is the observation set for task $\mathcal{M}_i$. Using this posterior distribution, an approximate MDP $\hat{M}_i$ is learned by leveraging previous tasks to approximate $\mathcal{M}_i$. This addition of the class layer makes the model hierarchical. The intuition is that the knowledge in a class can be transferred to an MDP task within the same class, but not to an MDP task outside the class.

To take into account the unknown number of MDP tasks in lifelong RL, a nonparametric infinite mixture model was used in the class layer. In the nonparametric infinite mixture model, it is assumed that there is an infinite number of classes (or mixture components), which account for the case of seeing a new MDP task that is dissimilar to all previous ones. Specifically, the Dirichlet process is applied. Dirichlet process is a stochastic process involving a base distribution G_0 and a positive scaling parameter α. The parameter α governs the probability with which the Dirichlet Process assumes a new class should be assigned. This new class is also called an *auxiliary class*. Using the above process, a Gibbs sampling process can be designed to repeatedly sample class assignments until convergence.

9.2.3   MTRL ALGORITHM

We now present the MTRL algorithm (see Algorithm 9.1). At the beginning, without having any MDP task, the hierarchical model parameters Ψ are initialized to uninformed values (line 1). When each new MDP task $\mathcal{M}_i$ arrives (line 2), the algorithm goes through two steps: (1) it applies the knowledge Ψ learned from the previous MDP tasks to learn an approximate MDP $\hat{M}_i$ for $\mathcal{M}_i$ (lines 4–10) and (2) it updates the old knowledge to generate the new knowledge from $\hat{M}_1, \ldots, \hat{M}_i$ (line 12) after considering the new task.

Algorithm 9.1 Hierarchical Bayesian MTRL Algorithm

Input: A sequence of MDP tasks $\mathcal{M}_1, \mathcal{M}_2, \ldots$
Output: Hierarchical model parameters Ψ

1: Initialize the hierarchical model parameters Ψ
2: **for** each MDP task $\mathcal{M}_i$ from $i = 1, 2, \ldots$ **do**
3:   // Step 1: apply the past knowledge for fast learning of the new MDP task $\mathcal{M}_i$
4:   $\mathcal{O}_i = \emptyset$; // $\mathcal{O}_i$ is the observation set for the environment in $\mathcal{M}_i$
5:   **while** policy π_i has not converged **do**
6:     $\hat{M}_i \leftarrow \text{SampleAnMDP}(P(M|\Psi, \mathcal{O}_i))$ // $P(M|\Psi, \mathcal{O}_i)$ is the posterior distribution
7:     $\pi_i = \text{Solve}(\hat{M}_i)$ // e.g., by value iteration
8:     Run π_i in $\mathcal{M}_i$ for k steps
9:     $\mathcal{O}_i = \mathcal{O}_i \cup \{\text{observations from } k \text{ steps}\}$
10:   **end while**
11:   // Step 2: learn the new parameters (knowledge) from $\hat{M}_1, \ldots, \hat{M}_i$
12:   $\Psi \leftarrow \text{UpdateModelParameters}(\Psi|\hat{M}_1, \ldots, \hat{M}_i)$
13: **end for**

For step 1, the function SampleAnMDP samples a set of MDPs based on the posterior distribution $P(M|\Psi, \mathcal{O}_i)$, where M denotes a random variable over all MDPs, and returns an MDP with the highest probability (say $\hat{M}_i$). This is how the past knowledge is used. We will explain this function in Section 9.2.5. An optimal policy π_i is then learned for $\hat{M}_i$ (line 7) using a method like value iteration [Sutton and Barto, 1998]. After π_i is obtained, it is applied for k steps in the $\mathcal{M}_i$ environment (line 8). This part is similar to Thompson sampling [Strens, 2000, Thompson, 1933, Wang et al., 2005] except that a set of MDPs is sampled first and the one with the highest probability is selected. The observations gathered from the k steps are added into the observation set $\mathcal{O}_i$, which changes the posterior distribution $P(M|\Psi, \mathcal{O}_i)$. The system then goes to the next iteration to sample a new $\hat{M}_i$. This process is repeated until the policy π_i converges.

For step 2, line 12 learns a new set of hierarchical model parameters Ψ from $\hat{M}_1, \ldots, \hat{M}_i$, which contains the class assignment for each MDP task and the model parameters associated with each class. Note that the function UpdateModelParameters (see Section 9.2.4) can automatically decide the number of classes, as well as the inherent class structure in the hierarchical model.

9.2.4   UPDATING HIERARCHICAL MODEL PARAMETERS

We first describe how to update the hierarchical model parameters Ψ (line 12 in Algorithm 9.1). Details about sampling of an MDP (line 6) will be discussed in the next subsection. Gibbs sampling is used to find the proper set of model parameters (see Algorithm 9.2). The techniques in Algorithm 9.2 can handle the situation where the base distribution G_0 is not conjugate to the component distribution. In Gibbs sampling, the Markov chain state includes Θ and $\mathcal{C}$, where $\Theta = \{\theta_1, \ldots, \theta_K\}$ (K is the number of existing classes) is the set of class parameters and $\mathcal{C} = \{C_1, \ldots, C_i\}$ is the set of class assignments. The use of auxiliary classes allows for the assignments of novel or new classes. m is the number of such auxiliary classes and is empirically set to a small value.

Algorithm 9.2 Update Hierarchical Model Parameters

Input: Model estimates $\{\hat{M}_1, \ldots, \hat{M}_i\}$ for MDP tasks $\{\mathcal{M}_1, \ldots, \mathcal{M}_i\}$, MDP distribution F given a class, Dirichlet Process $DP(G_0, \alpha)$
Output: Updated hierarchical model parameters $\hat{\Psi}$

1: Let i be the total number of MDPs seen so far.
2: Let m be the number of auxiliary classes
3: Initialize the Markov chain state $(\Theta_0, \mathcal{C}_0)$
4: $k \leftarrow 0$
5: **while** Gibbs sampling is not converged **do**
6:     $K = |\Theta_k|$
7:     **for** $c = K + 1$ **to** $K + m$ **do**
8:         Draw θ_c from G_0
9:         $\Theta_k = \Theta_k \cup \{\theta_c\}$
10:     **end for**
11:     $\hat{\Psi} = \{\Theta_k, \mathcal{C}_k\}$
12:     **for** $j = 1$ **to** i **do**
13:         $c_j = \text{SamplingClassAssignment}(\hat{\Psi}, \hat{M}_j, F, K, m, G_0, \alpha)$
14:     **end for**
15:     Remove all classes with zero MDPs
16:     $\Theta_{k+1} = \text{Sample}(P(\Theta_k | c_1, \ldots, c_i))$
17:     $\mathcal{C}_k = \{c_1, \ldots, c_i\}$
18:     $k \leftarrow k + 1$
19: **end while**
20: return $\hat{\Psi} = \{\Theta_k, \mathcal{C}_k\}$

In Algorithm 9.2, the Markov chain state is initialized with the current parameters (line 3). Lines 7–10 draw the parameters for each auxiliary class. Lines 12–14 call Algorithm 9.3 to sample a class assignment for each $\hat{M}_j$. Given the class assignments, a new set of class parameters are sampled (line 16). The sampling depends on the specific form of MDP distribution, and was not specified in the paper. After the burn-in period, Gibbs sampler keeps running until it converges. The final Markov state is returned to update the hierarchical model parameters (line 20).

9.2.5   SAMPLING AN MDP

Finally, we describe the function SampleAnMDP (line 6 in Algorithm 9.1), which samples an MDP. For accurate sampling, the agent or system needs to have an accurate hierarchical model. Then it should update its model parameters Ψ (knowledge) whenever a new observation is available. However, this is computationally expensive for LL, considering that the number of observations and the number of MDP tasks can both be large. Instead, Wilson et al. [2007] proposed to keep the parameters Ψ fixed when learning a new MDP. That's why line 12 in Algorithm 9.1 is outside of the while loop (lines 5–10). Note that Ψ includes the class assignments $\mathcal{C}$ and class parameters Θ, which together is called an informed prior, and they remain fixed during the exploration of a new MDP.

The process of generating an MDP is such that a class c is sampled first and the MDP $\hat{M}_i$ is sampled afterward based on the class. The class is sampled with the help of Algorithm 9.3. Here is how the past knowledge is used to help future learning. That is, if c belongs to a known class $c \in \{1, \ldots, K\}$, then the information in θ_c is used as the prior knowledge for exploration (see below). Otherwise, the agent uses a new class and samples the class parameters θ_c from the prior G_0 (no past knowledge is used).

SampleAnMDP works as follow: at the beginning, $\hat{M}_i$ is initialized by sampling from the informed prior, and C_i is initialized similarly. In subsequent iterations, after each set of observations (line 8 in Algorithm 9.1), the agent samples a sequence of class assignments C_i by running Algorithm 9.3 multiple times and picks the most probable one as the class assignment for $\hat{M}_i$. Recall that α controls how likely the returned class c is an auxiliary class (unseen class), i.e., $K + 1 \leq c \leq K + m$ (line 4 of Algorithm 9.3). Once the class c is sampled, the agent then samples an MDP from class c using the posterior distribution $P(M_i|\theta_c, \mathcal{O}_i)$. The algorithm is generic and applicable to different forms of MDP distribution F which lead to distinct specific sampling procedures. See the original paper, Wilson et al. [2007], for additional details.

9.3   PG-ELLA: LIFELONG POLICY GRADIENT REINFORCEMENT LEARNING

Instead of augmenting the stochastic gradient method with LL capability in Section 9.1, Bou Ammar et al. [2014] employed a policy gradient method [Sutton et al., 2000]. Specifically, Bou Ammar et al. [2014] extended a single-task policy gradient algorithm to an LL algorithm called *Policy Gradient Efficient Lifelong Learning Algorithm* (PG-ELLA). The lifelong idea in PG-ELLA is similar to that in ELLA [Ruvolo and Eaton, 2013b] (Section 3.4). In this section, we first introduce policy gradient RL and then present the PG-ELLA algorithm. Throughout this section, we adopt the notations in Bou Ammar et al. [2014].

Algorithm 9.3 Sampling Class Assignment

Input: Hierarchical model parameters Ψ, MDP parameter estimate $\hat{M}_j$, MDP distribution F given a class, the number of existing classes K, the number of auxiliary classes m, Dirichlet Process $DP(G_0, \alpha)$
Output: Class assignment C_j for $\hat{M}_j$

1: Let i be the total number of MDPs seen so far
2: Let $n_{-j,c}$ be the number of MDPs assigned to class c without considering class assignment of $\hat{M}_j$
3: Let $F_{c,j}$ denotes $F(\theta_c, \hat{M}_j)$, the probability of $\hat{M}_j$ in class c (the exact form may differ in different problems)
4: Sample and return C_j according to:

$$P(C_j = c) \propto \begin{cases} \frac{n_{-j,c}}{i-1+\alpha} F_{c,j}, & 1 \le c \le K \\ \frac{\alpha/m}{i-1+\alpha} F_{c,j}, & K+1 \le c \le K+m \end{cases}$$

9.3.1 POLICY GRADIENT REINFORCEMENT LEARNING

In RL, an agent sequentially chooses actions to perform to maximize its expected reward or return. As mentioned earlier, such problems are typically formalized as a Markov Decision Process (MDP) $\langle \mathcal{X}, \mathcal{A}, P, R, \lambda \rangle$. $\mathcal{X} \subseteq \mathbb{R}^d$ is the set of states that is potentially infinite with d being the dimension of the environment. $\mathcal{A} \subseteq \mathbb{R}^{d_a}$ is the set of all possible actions and d_a is the number of possible actions. $P : \mathcal{X} \times \mathcal{A} \times \mathcal{X} \to [0,1]$ is the state transition probability function, i.e., given a state and an action, it gives the probability of the next state. $R : \mathcal{X} \times \mathcal{A} \to \mathbb{R}$ is the reward function that provides the agent feedback. $\lambda \in [0,1)$ is the degree to which rewards are discounted over time.

At each time step h, being in the state $x_h \in \mathcal{X}$, the agent must choose an action $a_h \in \mathcal{A}$. After the action is taken, the agent transits to a new state $x_{h+1} \sim p(x_{h+1}|x_h, a_h)$ as given by P. At the same time, a reward $r_{h+1} = R(x_h, a_h)$ is sent to the agent as feedback. A *policy* is defined as a probability distribution over pairs of state and action, $\pi : \mathcal{X} \times \mathcal{A} \to [0,1]$. $\pi(a|x)$ indicates the probability of choosing action a given state x. The goal of RL is to find an optimal policy π^* that maximizes the expected return for the agent. The actual sequence of state-action pairs forms a *trajectory* $\tau = [x_{0:H}, a_{0:H}]$ over a possibly infinite horizon H.

Policy gradient methods have been widely applied in solving high-dimensional RL problems [Bou Ammar et al., 2014, Peters and Schaal, 2006, Peters and Bagnell, 2011, Sutton et al., 2000]. In a policy gradient method, the policy is represented by a parametric probability distribution $\pi_\theta(a|x) = p(a|x; \theta)$ that stochastically chooses action a given state x based on a vector θ of control parameters. The objective is to find the optimal parameters θ^* that maximize the expected average return:

$$\mathcal{J}(\boldsymbol{\theta}) = \int_{\mathbb{T}} p_{\boldsymbol{\theta}}(\tau)\mathfrak{R}(\tau)d\tau \ , \tag{9.3}$$

where $\mathbb{T}$ denotes the set of all possible trajectories. The distribution over the trajectory τ is defined as:

$$p_{\boldsymbol{\theta}}(\tau) = P_0(x_0) \prod_{h=0}^{H-1} p(x_{h+1}|x_h, a_h)\pi_{\boldsymbol{\theta}}(a_h|x_h) \ . \tag{9.4}$$

Here $P_0(x_0)$ represents the probability of the initial state. The average return $\mathfrak{R}(\tau)$ is defined as:

$$\mathfrak{R}(\tau) = \frac{1}{H}\sum_{h=0}^{H-1} r_{h+1} \ . \tag{9.5}$$

Most policy gradient algorithms learn the parameters $\boldsymbol{\theta}$ by maximizing a lower bound on the expected return of $\mathcal{J}(\boldsymbol{\theta})$ (Equation (9.3)). It compares the result of the current policy $\pi_{\boldsymbol{\theta}}$ and that of a new policy $\pi_{\tilde{\boldsymbol{\theta}}}$. As in Kober and Peters [2011], this lower bound can be obtained using Jensen's inequality and the concavity of the logarithm:

$$
\begin{aligned}
\log \mathcal{J}\left(\tilde{\boldsymbol{\theta}}\right) &= \log \int_{\mathbb{T}} p_{\tilde{\boldsymbol{\theta}}}(\tau)\mathfrak{R}(\tau)d\tau \\
&= \log \int_{\mathbb{T}} \frac{p_{\boldsymbol{\theta}}(\tau)}{p_{\boldsymbol{\theta}}(\tau)} p_{\tilde{\boldsymbol{\theta}}}(\tau)\mathfrak{R}(\tau)d\tau \\
&\geq \int_{\mathbb{T}} p_{\boldsymbol{\theta}}(\tau)\mathfrak{R}(\tau)\log\frac{p_{\tilde{\boldsymbol{\theta}}}(\tau)}{p_{\boldsymbol{\theta}}(\tau)}d\tau \quad \text{(using Jensen's inequality)} \\
&= -\int_{\mathbb{T}} p_{\boldsymbol{\theta}}(\tau)\mathfrak{R}(\tau)\log\frac{p_{\boldsymbol{\theta}}(\tau)}{p_{\tilde{\boldsymbol{\theta}}}(\tau)}d\tau \\
&\propto -\mathfrak{D}_{KL}\left(p_{\boldsymbol{\theta}}(\tau)\mathfrak{R}(\tau)||p_{\tilde{\boldsymbol{\theta}}}(\tau)\right) = \mathcal{J}_{\mathcal{L},\boldsymbol{\theta}}(\tilde{\boldsymbol{\theta}}) \ ,
\end{aligned}
\tag{9.6}
$$

where $\mathfrak{D}_{KL}$ denotes the KL-Divergence. From the above, one can minimize the KL-Divergence between the trajectory distribution $p_{\boldsymbol{\theta}}$ of the current policy $\pi_{\boldsymbol{\theta}}$ times its reward function $\mathfrak{R}$ and the trajectory distribution $p_{\tilde{\boldsymbol{\theta}}}$ of the new policy $\pi_{\tilde{\boldsymbol{\theta}}}$.

9.3.2 POLICY GRADIENT LIFELONG LEARNING SETTING

The problem setting of policy gradient LL is similar to the problem setting of ELLA (Efficient Lifelong Learning Algorithm) (Section 3.4.1). That is, the reinforcement learning tasks arrive sequentially in a lifelong manner. Each task t is an MDP $\langle \mathcal{X}^t, \mathcal{A}^t, P^t, R^t, \lambda^t \rangle$ with the initial state distribution P_0^t. Different from the supervised learning, each reinforcement learning task does not contain labeled training data. In each task, the agent learns multiple trajectories before moving to the next task. Let N be the number of tasks encountered so far. N may be unknown to the agent. The goal is to learn a set of *optimal* policies $\{\pi_{\boldsymbol{\theta}^1}^*, \pi_{\boldsymbol{\theta}^2}^*, \ldots, \pi_{\boldsymbol{\theta}^N}^*\}$ with corresponding parameters $\{\boldsymbol{\theta}^{1*}, \boldsymbol{\theta}^{2*}, \ldots, \boldsymbol{\theta}^{N*}\}$.

9.3.3   OBJECTIVE FUNCTION AND OPTIMIZATION

Similar to ELLA, PG-ELLA also assumes that each task model's parameters $\boldsymbol{\theta}^t$ can be represented by a linear combination of a set of shared latent components $\mathbf{L}$ (shared knowledge) and a task-specific coefficient vector $\mathbf{s}^t$, i.e., $\boldsymbol{\theta}^t = \mathbf{L}\mathbf{s}^t$ [Bou Ammar et al., 2014]. In other words, PG-ELLA maintains k sparsely shared basis model components for all task models. The k basis model components are represented by $\mathbf{L} \subseteq \mathbb{R}^{d \times k}$, where d is the model parameter dimension. The task-specific vector $\mathbf{s}^t$ should be sparse in order to accommodate the differences among tasks. The objective function of PG-ELLA is as follows:

$$\frac{1}{N} \sum_{t=1}^{N} \min_{\mathbf{s}^t} \left\{ -\mathcal{J}\left(\boldsymbol{\theta}^t\right) + \mu \|\mathbf{s}^t\|_1 \right\} + \lambda \|\mathbf{L}\|_F^2 \quad, \tag{9.7}$$

where $\| \cdot \|_1$ is the L_1 norm, which is controlled by μ as a convex approximation to the true vector sparsity. $\|\mathbf{L}\|_F^2$ is the Frobenius norm of matrix $\mathbf{L}$, and λ is the regularization coefficient for matrix $\mathbf{L}$. This objective function is closely related to Equation (3.4) in ELLA. This objective function is not jointly convex in $\mathbf{L}$ and $\mathbf{s}^t$. Thus, the alternating optimization strategy was adopted to find a local minimum, i.e., optimizing $\mathbf{L}$ while fixing $\mathbf{s}^t$ and optimizing $\mathbf{s}^t$ while fixing $\mathbf{L}$.

Combining Equations (9.6) and (9.7), we obtain the objective function below:

$$\frac{1}{N} \sum_{t=1}^{N} \min_{\mathbf{s}^t} \left\{ -\mathcal{J}_{\mathcal{L},\boldsymbol{\theta}}\left(\tilde{\boldsymbol{\theta}}^t\right) + \mu \|\mathbf{s}^t\|_1 \right\} + \lambda \|\mathbf{L}\|_F^2 \quad. \tag{9.8}$$

Note the following for $\mathcal{J}_{\mathcal{L},\boldsymbol{\theta}}\left(\tilde{\boldsymbol{\theta}}^t\right)$:

$$\mathcal{J}_{\mathcal{L},\boldsymbol{\theta}}\left(\tilde{\boldsymbol{\theta}}^t\right) \propto - \int_{\tau \in \mathbb{T}^t} p_{\boldsymbol{\theta}^t}(\tau) \mathfrak{R}^t(\tau) \log \left(\frac{p_{\boldsymbol{\theta}^t}(\tau) \mathfrak{R}^t(\tau)}{p_{\tilde{\boldsymbol{\theta}}^t}(\tau)} \right) d\tau \quad. \tag{9.9}$$

So the objective function can be rewritten as:

$$\frac{1}{N} \sum_{t=1}^{N} \min_{\mathbf{s}^t} \left\{ \left[\int_{\tau \in \mathbb{T}^t} p_{\boldsymbol{\theta}^t}(\tau) \mathfrak{R}^t(\tau) \log \left(\frac{p_{\boldsymbol{\theta}^t}(\tau) \mathfrak{R}^t(\tau)}{p_{\tilde{\boldsymbol{\theta}}^t}(\tau)} \right) d\tau \right] + \mu \|\mathbf{s}^t\|_1 \right\} + \lambda \|\mathbf{L}\|_F^2 \quad. \tag{9.10}$$

Again similar to ELLA, there are two major inefficiencies when solving the objective function: (a) the explicit dependence of *all* available trajectories of all tasks, and (b) the evaluation of a single candidate $\mathbf{L}$ depends on the optimization of $\mathbf{s}^t$ for each task t. To address the first issue, the second-order Taylor approximation is used to approximate the objective function. Following the steps in Section 3.4.3, one can yield the approximate objective function below:

$$\frac{1}{N} \sum_{t=1}^{N} \min_{\mathbf{s}^t} \left\{ \|\hat{\boldsymbol{\theta}}^t - \mathbf{L}\mathbf{s}^t\|_{\boldsymbol{H}^t}^2 + \mu \|\mathbf{s}^t\|_1 \right\} + \lambda \|\mathbf{L}\|_F^2 \quad, \tag{9.11}$$

$$H^t = \frac{1}{2}\nabla^2_{\tilde{\boldsymbol{\theta}}^t, \tilde{\boldsymbol{\theta}}^t} \left\{ \int_{\tau \in \mathbb{T}^t} p_{\boldsymbol{\theta}^t}(\tau)\Re^t(\tau) \log\left(\frac{p_{\boldsymbol{\theta}^t}(\tau)\Re^t(\tau)}{p_{\tilde{\boldsymbol{\theta}}^t}(\tau)}\right) d\tau \right\}\Bigg|_{\tilde{\boldsymbol{\theta}}^t = \hat{\boldsymbol{\theta}}^t} \quad \text{and}$$

$$\hat{\boldsymbol{\theta}}^t = \underset{\tilde{\boldsymbol{\theta}}^t}{\mathrm{argmin}} \left\{ \int_{\tau \in \mathbb{T}^t} p_{\boldsymbol{\theta}^t}(\tau)\Re^t(\tau) \log\left(\frac{p_{\boldsymbol{\theta}^t}(\tau)\Re^t(\tau)}{p_{\tilde{\boldsymbol{\theta}}^t}(\tau)}\right) d\tau \right\} .$$

The second issue arises when computing the objective function for a single $\mathbf{L}$. For each single candidate $\mathbf{L}$, an optimization problem must be solved to recompute each of the $\mathbf{s}^t$'s. When the number of tasks become large, this procedure becomes very expensive. The approach to remedying this issue follows that in Section 3.4.4. When task t is encountered, only $\mathbf{s}^t$ is updated while $\mathbf{s}^{t'}$ for all other tasks t' remain the same. Consequently, any changes to $\boldsymbol{\theta}^t$ will be transferred to other tasks only through the shared base $\mathbf{L}$. Ruvolo and Eaton [2013b] showed that this strategy does not significantly affect the quality of model fit when there are a large number of tasks. Using the previously computed values of $\mathbf{s}^t$, the following optimizing process is performed:

$$\mathbf{s}^t \leftarrow \underset{\mathbf{s}^t}{\mathrm{argmin}} \|\hat{\boldsymbol{\theta}}^t - \mathbf{L}_m \mathbf{s}^t\|^2_{\boldsymbol{H}^t} + \mu\|\mathbf{s}^t\|_1 , \text{ with fixed } \mathbf{L}_m, \text{ and}$$

$$\mathbf{L}_{m+1} \leftarrow \underset{\mathbf{L}}{\mathrm{argmin}} \frac{1}{N}\sum_{t=1}^{N}\left(\|\hat{\boldsymbol{\theta}}^t - \mathbf{L}\mathbf{s}^t\|^2_{\boldsymbol{H}^t} + \mu\|\mathbf{s}^t\|_1\right) + \lambda\|\mathbf{L}\|^2_F, \text{ with fixed } \mathbf{s}^t ,$$

where $\mathbf{L}_m$ refers to the value of the latent components at the mth iteration and t is assumed to be the particular task that the agent is working on. Additional details can be found in Bou Ammar et al. [2014].

9.3.4   SAFE POLICY SEARCH FOR LIFELONG LEARNING

PG-ELLA employs unconstrained optimization in learning. However, such unconstrained optimization could be fragile since the agent may learn to perform dangerous actions and cause physical damage to the agent or environment. Based on PG-ELLA, Bou Ammar et al. [2015c] proposed a safe lifelong learner for policy gradient reinforcement learning using an adversarial framework. It considered the safety constraints on each task when optimizing the overall performance. The objective function in Bou Ammar et al. [2015c] is:

$$\min_{\mathbf{L}, \mathbf{s}^t}\left[\eta^t \times l^t(\mathbf{L}\mathbf{s}^t)\right] + \mu\|\mathbf{s}^t\|_1 + \lambda\|\mathbf{L}\|^2_F \tag{9.12}$$

$$\begin{aligned} s.t. \ & \mathbf{A}^t\mathbf{L}\mathbf{s}^t \leq \mathbf{b}^t \quad \forall t \in \{1, 2, \ldots, N\} \\ & \lambda_{min}(\mathbf{L}\mathbf{L}^\mathrm{T}) \geq p \text{ and } \lambda_{max}(\mathbf{L}\mathbf{L}^\mathrm{T}) \leq q , \end{aligned}$$

where constraints $\mathbf{A}^t \in \mathbb{R}^{d \times d}$, where d is the model parameter dimension, and $\mathbf{b}^t \in \mathbb{R}^d$ represent the allowed policy combinations. λ_{min} and λ_{max} are the minimum and maximum eigenvalues. p and q are bounding constraints on Frobenius norm to ensure the shared knowledge is effective and safe to use. η^t is the design weight for each task. The above objective function aims to make sure that the knowledge is safely transferred across tasks and avoid causing the agent to learn and perform irrational actions. For the method used in solving the optimization problem, please refer to Bou Ammar et al. [2015c].

9.3.5 CROSS-DOMAIN LIFELONG REINFORCEMENT LEARNING

The works in Bou Ammar et al. [2014, 2015c] above assume that the tasks come from a single task domain, i.e., they share a common state and action space. When the tasks have different state and/or action spaces, an *inter-task mapping* [Taylor et al., 2007] is usually needed to serve as a bridge between tasks. Taylor et al. [2007] studied transfer learning for reinforcement learning in this setting, i.e., transferring from one source domain to one target domain where the two domains have different state and action spaces. Given that an inter-task mapping is provided to an agent as input, Taylor et al. [2007] showed that the agent can learn one task and then significantly reduce the time it takes to learn another task.

Also in the transfer learning setting, Bou Ammar et al. [2015b] proposed an algorithm to automatically discover the inter-task mapping between two tasks. They focused on constructing an inter-state mapping and demonstrated the effectiveness of applying it from one task to another. The proposed method contains two steps. (1) It learns an inter-state mapping using the Unsupervised Manifold Alignment (UMA) method in Wang and Mahadevan [2009]. In particular, two sets of trajectories of states are collected from the source task and target task, respectively. Then, each set is transformed to a state feature vector on which UMA is applied. (2) Given the learned inter-state mapping, a set of initial states in the target task is mapped into the states in the source task. Then, based on the mapped source task states, the optimal source task policy is used to produce a set of optimal state trajectories. Such optimal state trajectories are then mapped back to the target task to generate target task-specific trajectories. However, the work in Wang and Mahadevan [2009] does not generalize well to the LL scenario as it only learns the mapping between a pair of tasks. It is computationally expensive to learn the mapping between each pair of tasks from a large pool of tasks for LL. Isele et al. [2016] also proposed a zero-shot LL method that models the inter-task relationship via task descriptors.

Bou Ammar et al. [2015a] proposed a more efficient method to maintain and transfer knowledge in a sequence of tasks in the lifelong setting. It is closely related to PG-ELLA [Bou Ammar et al., 2014]. The difference is that Bou Ammar et al. [2015a] allows the tasks to come from different domains, i.e., from different state and/or action spaces. Bou Ammar et al. [2015a] assumed that all tasks can be grouped into different task groups where tasks within a task group are assumed to share a common state and action space. Formally, instead of formulating the task parameters as $\theta^t = \mathbf{L}\mathbf{s}^t$ as in PG-ELLA (Section 9.3.3), Bou Ammar et al. [2015a] formulated them as $\theta^t = B^{(g)}\mathbf{s}^t$ where g is the task group of $B^{(g)}$, which is the latent model components shared within g. Similar to PG-ELLA, $\mathbf{s}^t$ is assumed to be sparse to accommodate distinct tasks. Furthermore, $B^{(g)}$ is assumed to be $\Psi^{(g)}\mathbf{L}$ where $\mathbf{L}$ is the global latent model components (same as that in PG-ELLA), and $\Psi^{(g)}$ maps the shared latent components $\mathbf{L}$ into the basis for each group g of tasks. Basically, Bou Ammar et al. [2015a] added another layer, i.e., task group, to model the tasks from different domains. Theoretical guarantees were provided on the stability of the approach as the number of tasks and groups increases. Please refer to the original paper in Bou Ammar et al. [2015a] for additional details.

9.4   SUMMARY AND EVALUATION DATASETS

This chapter introduced the existing LL work in the context of RL. Again, the current work is not extensive. This is perhaps partly due to the fact that RL was not as popular as traditional supervised learning in the past because of fewer real-life applications. However, RL has come to the mainstream due to the AlphaGo's success in beating the best human player in the board game of Go [Silver et al., 2016, 2017]. Although games have been the traditional application area of RL, it has significantly more applications than just games. With the increased popularity of physical as well as software robots (such as chatbots and intelligent personal assistants) that need to interact with human beings and other robots in real-life environments, RL will become more and more important. Lifelong RL will be important too because it is very hard to collect a large number of training examples in such real-life interactive environments with each individual human person or robot. The system has to learn and accumulate knowledge from all possible environments that it has experiences in to adapt itself to a new environment quickly and to perform its task well.

Evaluation Datasets

Finally, to help researchers in the field, we summarize the evaluation datasets used in the papers discussed in this chapter. Tanaka and Yamamura [1997] used 9×9 mazes data in their evaluation. Wilson et al. [2007] tested their hypotheses using a synthetic colored maze data where the task is to go from one location to another following the least cost path. PG-ELLA [Bou Ammar et al., 2014] was evaluated on three benchmark dynamic systems: Simple Mass Spring Damper [Bou Ammar et al., 2014], Cart-Pole [Bocsi et al., 2013], and Three-Link Inverted Pendulum [Bou Ammar et al., 2014]. Simple Mass Spring Damper and Cart-Pole were also used in Bou Ammar et al. [2015c]. Other than the three dynamic systems, Quadrotor [Bouabdallah, 2007] was also used for evaluation in Bou Ammar et al. [2015b]. Bou Ammar et al. [2015a] additionally considered Bicycle and Helicopter systems. Tessler et al. [2017] used the environment State space.

CHAPTER 10

Conclusion and Future Directions

This book surveyed many existing ideas and techniques of lifelong (machine) learning (LL). It also briefly covered closely related learning paradigms such as transfer learning and multi-task learning (MTL), and discussed their differences from LL. There have been some confusions among researchers and practitioners about the differences between these learning paradigms, which is not surprising as they are indeed similar and related. Hopefully, our new definition of LL in Section 1.4 and subsequent discussions in Chapter 2 help clarify the differences and resolve the confusions.

Although LL was originally proposed in 1995, as mentioned in chapter 1, the research in the field has not been extensive due to many factors, e.g., its own difficulty, lack of big data in the past, and the emphasis of statistical and algorithmic learning in the machine learning (ML) community in the past two decades. However, with the resurgence of AI and the progress and maturity of statistical ML algorithms, LL is becoming increasingly important because the ultimate goal of ML is to learn continuously, interactively, and autonomously in diverse domains and in open environments to enable intelligent agents to become more and more knowledgeable and better and better at learning. Applications such as intelligent assistants, chatbots, self-driving cars, and other software and hardware robots all call for LL. A system is not intelligent in the general sense without the ability to learn many different types of knowledge, accumulate the knowledge over time, and use the knowledge to learn more and to learn better. Even if a system is extremely good at performing one difficult task, e.g., playing Go like AlphaGo or playing chess like Deep Blue, it is not an intelligent system in the general sense. Because of the physical limitations of human brains, our thinking, reasoning, and problem solving are probably not or cannot be optimized for complex tasks. A machine does not have these limitations and is bound to outperform human beings on well-defined and narrow tasks in restricted environments. However, this does not necessarily make the machine intelligent, at least not in the sense of the general human intelligence. Traditionally, we often equate intelligence to some special mental capabilities or talents because we compare humans with humans. However, intelligence is more about humans' baseline perceptual and cognitive capabilities, which enable them to continuously learn new knowledge about almost anything and to apply the knowledge seamlessly to solve all kinds of problems. This forms a type of virtuous circle.

We believe that now it is time to put a significant amount of effort in the research of LL for many reasons. First, there is a huge amount of data available now which enables a system to learn a large quantity of diverse knowledge. Without a large volume of existing knowledge, it is very difficult to learn more knowledge by leveraging the past knowledge. This is analogous to human learning. The more we know, the more and better we are able to learn. Second, statistical ML is becoming mature. Further improvements are becoming more and more difficult, while using the past learned knowledge to help learning is a natural way going forward, which aims to imitate the human learning process. Existing research has shown that LL is highly effective. Third, with the increased use of intelligent personal assistants, chatbots, and physical robots that interact with humans and other systems in real-life and open environments, continuous LL capabilities are becoming increasingly necessary. We expect a large amount of research will appear in the near future, which may result in major breakthroughs.

Below, we would like to highlight some challenging problems and future directions to encourage more research in LL. Their solutions can have fundamental impact on LL specifically and on ML and AI in general.

1. **Correctness of knowledge**: How to know whether a piece of past knowledge is correct is crucial for LL. Because LL leverages the past knowledge to help future learning, incorrect past knowledge can be very harmful. In a nutshell, LL is a continuous bootstrapping learning process. Errors can propagate from previous tasks to subsequent tasks and result in more and more errors. This problem must be solved or mitigated to a great extent to ensure that LL is effective. Human beings solve this problem quite effectively. Even if mistakes are made initially, they can correct them later if new evidences appear. They can also backtrack and fix the errors along with the wrong inferences made based on the errors. An LL system should be able to do the same. Some existing LL systems have already tried to address this problem. For example, Chen and Liu [2014a] used frequent pattern mining to find those must-links (past knowledge) that appear in multiple domains and assumed those frequent must-links are more likely to be correct. They also explicitly checked the validity of the past knowledge in the modeling process. The NELL system [Mitchell et al., 2015] deals with the problem by ensuring that the same item is extracted from multiple sources, using multiple strategies or meeting some type constraints. However, the existing methods are still quite rudimentary. Their recalls are low and can still get wrong knowledge.

2. **Applicability of knowledge**: How to know whether a piece of knowledge is applicable to a new learning task is also critical for LL. Although a piece of knowledge may be correct and applicable in the context of some previous tasks, it may not be applicable to the current task due to the wrong context. Without solving this problem, LL will not be effective either. Again, the systems in Chen and Liu [2014a,b], Chen et al. [2015], and Shu et al. [2016] have proposed some preliminary mechanisms to deal with the problem in the contexts of topic modeling, supervised classification, and belief propagation. However, the problem is

far from being solved as these solutions are still specific to specific problems. No general methods have yet been proposed. Much further research is needed. Clearly, this and the above problem are closely related.

3. **Knowledge representation and reasoning**: In the early days of AI, a significant amount of research was done on logic-based knowledge representation and reasoning. But in the past 20 years, AI research has shifted focus to statistical ML based on optimization. Since LL has a knowledge base (KB), knowledge representation and reasoning are naturally relevant and important. Reasoning allows the system to infer new knowledge from existing knowledge, which can be used in the new task learning. Important questions to be answered include what forms of knowledge are important, how to represent them, and what kinds of reasoning capabilities are useful to LL. So far, little research has been done to address these questions in the context of LL. Knowledge in existing LL systems is mainly represented based on the direct needs of the specific learning algorithms or applications. They still do not have the reasoning ability, except NELL [Mitchell et al., 2015], which has some limited reasoning capability.

4. **Learning with tasks of multiple types and/or from different domains**: Much of the current research of LL focuses on multiple tasks of the same type. In this case, it is easier to make use of the past knowledge. If different types of tasks are involved (e.g., entity recognition and attribute extraction), in order to transfer past knowledge from one type of task to another type, we need to make connections between these types of tasks. Otherwise, knowledge is hard to use across tasks. Again, the NELL system [Mitchell et al., 2015] made some attempts to do this. Ideally, this can be done automatically, but it is hard because the connection needs to be made via some higher-level knowledge, which has to be learned separately.

   When the tasks are from different domains, LL is also more challenging as it is likely to need higher-level knowledge too to bridge the gap and to find the relatedness or similarity among the tasks [Bou Ammar et al., 2015a] in order to ensure knowledge applicability. In some cases, one may even need to learn from a large number of domains because each domain only contributes a tiny amount of knowledge (some domains may contribute none) that is useful to the new task [Chen and Liu, 2014a,b, Wang et al., 2016]. When multiple types of tasks from very different domains are all involved, the challenge will be even greater.

5. **Self-motivated learning**: Current ML techniques typically require human users to give learning tasks and to provide a large volume of training data (except in a few cases where the agent can learn by interacting with a simulator). If a robot is to interact with its real-world environment and learn continuously, it needs to identify and formulate its own learning tasks and collect its own training data in its exploration of the world. For example, if a robot sees a person that it has never seen before, it should take a video or

many pictures of the person to collect positive training data. Actually, in this case, recognizing a stranger itself is already a challenge. It needs open-world learning (Chapter 5) [Fei et al., 2016], which most current supervised learning algorithms cannot do because they make the closed-world assumption that only those classes that appeared in training can appear in testing. In practice, this assumption is often violated. Another example is human-machine conversation. Future chatbots must be able to learn during conversation, extracting knowledge from user utterances and asking the human user when it does not understand something or encounters some new concepts (Chapter 8). We human beings do these all the time, which make us learn more and more in a self-motivated manner and become more and more knowledgeable. In more general terms, self-motivated learning means that the robot or the intelligent agent has a sense of curiosity and is interested in exploring the unknown and learning new things by itself in the exploration process. Clearly, this is closely related to unsupervised learning and reinforcement learning. These forms of learning and, for that matter, integrated learning of all learning forms should be made self-motivated. Note that self-motivated learning described here is different from self-taught learning or unsupervised feature learning reported in Raina et al. [2007]. In self-taught learning, a large amount of unlabeled data is used to learn a good feature representation of the input. The learned feature representation and a small amount of labeled data are then employed to build a classifier by applying a supervised learning method.

6. **Self-supervised learning**: In the traditional ML paradigm, a large volume of manually labeled training data is needed for accurate learning. However, it is impossible for humans to label everything in the world, which is way too complex, too many, and ever-changing. So for effective LL, in most situations the agent has to learn continuously by itself in a self-supervised manner by picking up implicit or explicit feedback or clues from humans or the environment to serve as the supervised information for it to learn without asking the humans to explicitly label the data. For example, in self-driving, if our autonomous car sees the car in front of it drive over a small pothole in the middle of the road, it can assume that the pothole is not dangerous. More generally, the car can learn from humans' driving behaviors through imitation learning, by listening to the user's verbal feedback or instructions, and even by asking the user questions to gain supervised information in a natural way. Additionally, knowledge learned previously from books and other authoritative sources and agent's own experiences can serve as supervised information too.

7. **Lifelong natural language learning**: Here we reiterate that NLP is perhaps one of the most suitable application areas for LL. First of all, most concepts are applicable across domains and tasks because the same words or phrases are used in different domains with the same or very similar meanings. Taking information extraction as an example, it is unclear whether the human brain has a complex algorithm like HMM or CRF for extraction, but human beings clearly can do so well in entity recognition. We believe that one of the key reasons is that when we are given a particular extraction or recognition task, we already

know most of the answers as we have learned and accumulated a great deal of entities in the past and know how to spot entities in the text from our past experiences. Second, all NLP tasks are closely related to each other as we discussed in Chapter 1, which is obvious because they together make the meaning of a sentence. Thus, the knowledge learned from one task can help learning of other tasks.

8. **Compositional learning**: Learning compositionally is likely to be very important for LL. Classic ML is not compositional. For example, as humans, we learn a language by learning individual words and phrases first, and then sentences, paragraphs, and full documents. The knowledge gained from this kind of learning is highly reusable. The current machine learning by labeling each entire sentence or even entire document with a single label is quite unnatural. The learned knowledge from such labeling is also hard to be reused. There are simply too many, almost an infinite number of possible sentences, which makes it very difficult to learn things that do not occur frequently. For statistical ML to work, the data must occur sufficiently frequently in order to compute reliable statistics. However, if it is possible to learn in a bottom-up fashion, from words, phrases, to sentences and whole documents, it is possible to understand those infrequent sentences because each of their component words or phrases may have appeared frequently. The syntactic structures of the sentences may have appeared frequently too. We believe that people learn compositionally. Compositional learning is especially useful for image recognition and natural language processing. For example, we not only can recognize a person as a whole, but also his/her face, head, arms, legs, torso, etc. For the head, you can recognize, eyes, mouth, nose, eyebrows, etc. Current learning algorithms do not learn compositionally. Compositional learning is likely to be very important for LL simply because it allows the system to share knowledge and to compose at any level of granularity.

This list of directions or challenging problems is by no means exhaustive. There are many other challenges too. As an emerging field, current LL methods and systems are still rudimentary. But the journey of 1,000 miles begins with the first step. The research area is a wide open field. A significant amount of research is still needed in order to make breakthroughs. Yet practical applications and intelligent systems call for this type of advanced ML in order to fundamentally advance the artificial intelligence research and applications. In the near future, we envisage that a number of large and complex learning systems will be built with the LL capability. Such systems with large KBs will enable major progress to be made. Without a great deal of prior knowledge already, it is difficult to learn more.

Bibliography

Wickliffe C. Abraham and Anthony Robins, (2005). Memory retention–the synaptic stability vs. plasticity dilemma. *Trends in Neurosciences*, 28(2):73–78. DOI: 10.1016/j.tins.2004.12.003. 55

Gediminas Adomavicius and Alexander Tuzhilin, (2005). Toward the next generation of recommender systems: A survey of the state-of-the-art and possible extensions. *IEEE Transactions on Knowledge and Data Engineering*, 17(6), pages 734–749. DOI: 10.1109/tkde.2005.99. 118

Rakesh Agrawal and Ramakrishnan Srikant, (1994). Fast algorithms for mining association rules. In *VLDB*, pages 487–499. 97, 121

Rahaf Aljundi, Punarjay Chakravarty, and Tinne Tuytelaars, (2016). Expert gate: Lifelong learning with a network of experts. *CoRR, abs/1611.06194*, 2. DOI: 10.1109/cvpr.2017.753. 57, 67, 68, 69, 74

Rahaf Aljundi, Francesca Babiloni, Mohamed Elhoseiny, Marcus Rohrbach, and Tinne Tuytelaars, (2017). Memory aware synapses: Learning what (not) to forget. *ArXiv Preprint ArXiv:1711.09601*. 58, 74

Naomi S. Altman, (1992). An introduction to kernel and nearest-neighbor nonparametric regression. *The American Statistician*, 46(3), pages 175–185. DOI: 10.2307/2685209. 37

David Ameixa, Luisa Coheur, Pedro Fialho, and Paulo Quaresma, (2014). Luke, I am your father: Dealing with out-of-domain requests by using movies subtitles. In *International Conference on Intelligent Virtual Agents*. DOI: 10.1007/978-3-319-09767-1_2. 131

Rie Kubota Ando and Tong Zhang, (2005). A high-performance semi-supervised learning method for text chunking. In *ACL*, pages 1–9. DOI: 10.3115/1219840.1219841. 21

Marcin Andrychowicz, Misha Denil, Sergio Gomez, Matthew W Hoffman, David Pfau, Tom Schaul, Brendan Shillingford, and Nando De Freitas. Learning to learn by gradient descent by gradient descent. In *NIPS*, pages 3981–3989, 2016. 33

David Andrzejewski, Xiaojin Zhu, and Mark Craven, (2009). Incorporating domain knowledge into topic modeling via Dirichlet forest priors. In *ICML*, pages 25–32. DOI: 10.1145/1553374.1553378. 92

David Andrzejewski, Xiaojin Zhu, Mark Craven, and Benjamin Recht, (2011). A framework for incorporating general domain knowledge into latent Dirichlet allocation using first-order logic. In *IJCAI*, pages 1171–1177. DOI: 10.5591/978-1-57735-516-8/IJCAI11-200. 92

Gabor Angeli, Melvin J. Premkumar, and Christopher D. Manning, (2015). Leveraging linguistic structure for open domain information extraction. In *ACL*. DOI: 10.3115/v1/p15-1034. 138

Bernard Ans, Stéphane Rousset, Robert M. French, and Serban Musca, (2004). Self-refreshing memory in artificial neural networks: Learning temporal sequences without catastrophic forgetting. *Connection Science*, 16(2):71–99. DOI: 10.1080/09540090412331271199. 58

Andreas Argyriou, Theodoros Evgeniou, and Massimiliano Pontil, (2008). Convex multi-task feature learning. *Machine Learning*, 73(3), pages 243–272. DOI: 10.1007/s10994-007-5040-8. 27

Rafael E. Banchs and Haizhou Li, (2012). Iris: A chat-oriented dialogue system based on the vector space model. In *Proc. of the ACL System Demonstrations*, pages 37–42. 131

Bikramjit Banerjee and Peter Stone, (2007). General game learning using knowledge transfer. In *IJCAI*, pages 672–677. 32

Michele Banko and Oren Etzioni, (2007). Strategies for lifelong knowledge extraction from the Web. In *K-CAP*, pages 95–102. DOI: 10.1145/1298406.1298425. 111

Jonathan Baxter, (2000). A model of inductive bias learning. *Journal of Artificial Intelligence Research*, 12, pages 149–198. 26

Shai Ben-David and Reba Schuller, (2003). Exploiting task relatedness for multiple task learning. In *COLT*. DOI: 10.1007/978-3-540-45167-9_41. 26

Abhijit Bendale and Terrance E Boult, (2015). Towards open world recognition. In *Proc. of the IEEE Conference on Computer Vision and Pattern Recognition*, pages 1893–1902. DOI: 10.1109/cvpr.2015.7298799. 7, 77, 78, 79

Abhijit Bendale and Terrance E. Boult, (2016). Towards open set deep networks. In *Proc. of the IEEE Conference on Computer Vision and Pattern Recognition*, pages 1563–1572. DOI: 10.1109/cvpr.2016.173. 79, 85, 86, 88

Yoshua Bengio, (2009). Learning deep architectures for AI. *Foundations and Trends {®} in Machine Learning*, 2(1), pages 1–127. DOI: 10.1561/2200000006. 24

Yoshua Bengio, (2012). Deep learning of representations for unsupervised and transfer learning. *Unsupervised and Transfer Learning Challenges in Machine Learning*, 7. 25

James Bergstra and Yoshua Bengio, (2012). Random search for hyper-parameter optimization. *Journal of Machine Learning Research*, 13(Feb):281–305. 72

Steffen Bickel, Michael Brückner, and Tobias Scheffer, (2007). Discriminative learning for differing training and test distributions. In *ICML*, pages 81–88. DOI: 10.1145/1273496.1273507. 21

David M. Blei, Andrew Y. Ng, and Michael I. Jordan, (2003). Latent Dirichlet allocation. *The Journal of Machine Learning Research*, 3, pages 993–1022. 91, 93

John Blitzer, Ryan McDonald, and Fernando Pereira, (2006). Domain adaptation with structural correspondence learning. In *EMNLP*, pages 120–128. DOI: 10.3115/1610075.1610094. 22

John Blitzer, Mark Dredze, and Fernando Pereira, (2007). Biographies, bollywood, boom-boxes and blenders: Domain adaptation for sentiment classification. In *ACL*, pages 440–447. 22, 72

Avrim Blum and Tom Mitchell, (1998). Combining labeled and unlabeled data with co-training. In *COLT*, pages 92–100. DOI: 10.1145/279943.279962. 24

Botond Bocsi, Lehel Csató, and Jan Peters, (2013). Alignment-based transfer learning for robot models. In *IJCNN*, pages 1–7. DOI: 10.1109/ijcnn.2013.6706721. 152

Danushka Bollegala, Takanori Maehara, and Ken-ichi Kawarabayashi, (2015). Unsupervised cross-domain word representation learning. In *ACL*. http://www.aclweb.org/anthology/P15-1071 DOI: 10.3115/v1/p15-1071. 52

Danushka Bollegala, Kohei Hayashi, and Ken-ichi Kawarabayashi, (2017). Think globally, embed locally—locally linear meta-embedding of words. *ArXiv*. 52

Edwin V. Bonilla, Kian M. Chai, and Christopher Williams, (2008). Multi-task Gaussian process prediction. In *NIPS*, pages 153–160. 22

Antoine Bordes, Seyda Ertekin, Jason Weston, and Léon Bottou, (2005). Fast Kernel classifiers with online and active learning. *The Journal of Machine Learning Research*, 6, pages 1579–1619. 31

Antoine Bordes, Jason Weston, Ronan Collobert, and Yoshua Bengio, (2011). Learning structured embeddings of knowledge bases. In *AAAI*. 132

Antoine Bordes, Nicolas Usunier, Alberto Garcia-Duran, Jason Weston, and Oksana Yakhnenko, (2013). Translating embeddings for modeling multi-relational data. In *NIPS*. 132

Haitham Bou Ammar, Eric Eaton, Jose Marcio Luna, and Paul Ruvolo, (2015a). Autonomous cross-domain knowledge transfer in lifelong policy gradient reinforcement learning. In *AAAI*. 8, 140, 151, 152, 155

Haitham Bou Ammar, Eric Eaton, Paul Ruvolo, and Matthew E. Taylor, (2014). Online multi-task learning for policy gradient methods. In *ICML*, pages 1206–1214. 8, 15, 140, 146, 147, 149, 150, 151, 152

Haitham Bou Ammar, Eric Eaton, Paul Ruvolo, and Matthew E. Taylor, (2015b). Unsupervised cross-domain transfer in policy gradient reinforcement learning via manifold alignment. In *AAAI*. 151, 152

Haitham Bou Ammar, Rasul Tutunov, and Eric Eaton, (2015c). Safe policy search for lifelong reinforcement learning with sublinear regret. In *ICML*. 8, 140, 150, 151, 152

Samir Bouabdallah, (2007). *Design and Control of Quadrotors with Application to Autonomous Flying*. Ph.D. thesis, Ecole Polytechnique Federale de Lausanne. DOI: 10.5075/epfl-thesis-3727. 152

Hervé Bourlard and Yves Kamp, (1988). Auto-association by multilayer perceptrons and singular value decomposition. *Biological Cybernetics*, 59(4–5):291–294. DOI: 10.1007/bf00332918. 68

Jordan L. Boyd-Graber, David M. Blei, and Xiaojin Zhu, (2007). A topic model for word sense disambiguation. In *EMNLP-CoNLL*, pages 1024–1033. 91

Greg Brockman, Vicki Cheung, Ludwig Pettersson, Jonas Schneider, John Schulman, Jie Tang, and Wojciech Zaremba, (2016). Openai gym. *ArXiv Preprint ArXiv:1606.01540*. 75

Emma Brunskill and Lihong Li, (2014). PAC-inspired option discovery in lifelong reinforcement learning. In *ICML*, pages 316–324. 140

Chris Buckley, Gerard Salton, and James Allan, (1994). The effect of adding relevance information in a relevance feedback environment. In *SIGIR*, pages 292–300. DOI: 10.1007/978-1-4471-2099-5_30. 84

Lucian Busoniu, Robert Babuska, Bart De Schutter, and Damien Ernst, (2010). *Reinforcement Learning and Dynamic Programming Using Function Approximators*, vol. 39. CRC press. DOI: 10.1201/9781439821091. 32

Raffaello Camoriano, Giulia Pasquale, Carlo Ciliberto, Lorenzo Natale, Lorenzo Rosasco, and Giorgio Metta, (2017). Incremental robot learning of new objects with fixed update time. In *IEEE International Conference on Robotics and Automation (ICRA)*, pages 3207–3214. 58 DOI: 10.1109/icra.2017.7989364.

Andrew Carlson, Justin Betteridge, and Bryan Kisiel, (2010a). Toward an architecture for never-ending language learning. In *AAAI*, pages 1306–1313. 8, 111, 114, 115

Andrew Carlson, Justin Betteridge, Richard C. Wang, Estevam R. Hruschka Jr., and Tom M. Mitchell, (2010b). Coupled semi-supervised learning for information extraction. In *WSDM*, pages 101–110. DOI: 10.1145/1718487.1718501. 115, 116

Rich Caruana, (1997). Multitask learning. *Machine Learning*, 28(1), pages 41–75. DOI: 10.1007/978-1-4615-5529-2_5. 10, 26, 38, 39, 53, 56

Chih-Chung Chang and Chih-Jen Lin, (2011). LIBSVM: A library for support vector machines. *ACM Transactions on Intelligent Systems and Technology (TIST)*, 2(3), page 27. DOI: 10.1145/1961189.1961199. 82

Jonathan Chang, Jordan Boyd-Graber, Wang Chong, Sean Gerrish, and David M. Blei, (2009). Reading tea leaves: How humans interpret topic models. In *NIPS*, pages 288–296. 92

Zhiyuan Chen and Bing Liu, (2014a). Topic modeling using topics from many domains, lifelong learning and big data. In *ICML*, pages 703–711. 7, 13, 14, 15, 16, 91, 92, 94, 98, 109, 154, 155

Zhiyuan Chen and Bing Liu, (2014b). Mining topics in documents: Standing on the shoulders of big data. In *KDD*, pages 1116–1125. DOI: 10.1145/2623330.2623622. 7, 13, 15, 16, 53, 89, 91, 92, 94, 100, 101, 102, 103, 104, 105, 106, 109, 154, 155

Zhiyuan Chen, Bing Liu, and M. Hsu, (2013a). Identifying intention posts in discussion forums. In *NAACL-HLT*, pages 1041–1050. 23, 24

Jianhui Chen, Lei Tang, Jun Liu, and Jieping Ye, (2009). A convex formulation for learning shared structures from multiple tasks. In *ICML*, pages 137–144. DOI: 10.1145/1553374.1553392. 10, 26

Jianhui Chen, Jiayu Zhou, and Jieping Ye, (2011). Integrating low-rank and group-sparse structures for robust multi-task learning. In *KDD*, pages 42–50. DOI: 10.1145/2020408.2020423. 27

Zhiyuan Chen, Arjun Mukherjee, and Bing Liu, (2014). Aspect extraction with automated prior knowledge learning. In *ACL*, pages 347–358. DOI: 10.3115/v1/p14-1033. 91, 94

Zhiyuan Chen, Nianzu Ma, and Bing Liu, (2015). Lifelong learning for sentiment classification. In *ACL (short paper)*, vol. 2, pages 750–756. DOI: 10.3115/v1/p15-2123. 7, 14, 15, 16, 36, 46, 47, 49, 53, 154

Zhiyuan Chen, Arjun Mukherjee, Bing Liu, Meichun Hsu, Malu Castellanos, and Riddhiman Ghosh, (2013b). Discovering coherent topics using general knowledge. In *CIKM*, pages 209–218. DOI: 10.1145/2505515.2505519. 92

Zhiyuan Chen, Arjun Mukherjee, Bing Liu, Meichun Hsu, Malu Castellanos, and Riddhiman Ghosh, (2013c). Exploiting domain knowledge in aspect extraction. In *EMNLP*, pages 1655–1667. 92, 105

Hao Cheng, Hao Fang, and Mari Ostendorf, (2015). Open-domain name error detection using a multi-task RNN. In *EMNLP*, pages 737–746. DOI: 10.18653/v1/d15-1085. 30

Kenneth Ward Church and Patrick Hanks, (1990). Word association norms, mutual information, and lexicography. *Computational Linguistics*, 16(1), pages 22–29. DOI: 10.3115/981623.981633. 98

Christopher Clingerman and Eric Eaton, (2017). Lifelong learning with Gaussian processes. In *Joint European Conference on Machine Learning and Knowledge Discovery in Databases*, pages 690–704, Springer. DOI: 10.1007/978-3-319-71246-8_42. 36

Gregory Cohen, Saeed Afshar, Jonathan Tapson, and André van Schaik, (2017). Emnist: An extension of mnist to handwritten letters. *ArXiv Preprint ArXiv:1702.05373*. DOI: 10.1109/ijcnn.2017.7966217. 88

Ronan Collobert and Jason Weston, (2008). A unified architecture for natural language processing: Deep neural networks with multitask learning. In *ICML*, pages 160–167. DOI: 10.1145/1390156.1390177. 30, 121

Ronan Collobert, Jason Weston, Léon Bottou, Michael Karlen, Koray Kavukcuoglu, and Pavel Kuksa, (2011). Natural language processing (almost) from scratch. *Journal of Machine Learning Research*, 12, pages 2493–2537. 30, 85

Robert Coop, Aaron Mishtal, and Itamar Arel, (2013). Ensemble learning in fixed expansion layer networks for mitigating catastrophic forgetting. *IEEE Transactions on Neural Networks and Learning Systems*, 24(10):1623–1634. DOI: 10.1109/tnnls.2013.2264952. 73

Mark Craven, Dan DiPasquo, Dayne Freitag, Andrew McCallum, Tom Mitchell, Kamal Nigam, and Seán Slattery, (1998). Learning to extract symbolic knowledge from the world wide web. In *AAAI*, pages 509–516. 111

Wenyuan Dai, Gui-Rong Xue, Qiang Yang, and Yong Yu, (2007a). Co-clustering based classification for out-of-domain documents. In *KDD*, pages 210–219. DOI: 10.1145/1281192.1281218. 22

Wenyuan Dai, Gui-rong Xue, Qiang Yang, and Yong Yu, (2007b). Transferring naive Bayes classifiers for text classification. In *AAAI*. 21, 23, 24

Wenyuan Dai, Qiang Yang, Gui-Rong Xue, and Yong Yu, (2007c). Boosting for transfer learning. In *ICML*, pages 193–200. DOI: 10.1145/1273496.1273521. 21

Cristian Danescu-Niculescu-Mizil and Lillian Lee, (2011). Chameleons in imagined conversations: A new approach to understanding coordination of linguistic style in dialogs. In *Proc. of the Workshop on Cognitive Modeling and Computational Linguistics, (ACL).* 138

Cristian Danescu-Niculescu-Mizil, Lillian Lee, Bo Pang, and Jon Kleinberg, (2012). Echoes of power: Language effects and power differences in social interaction. In *Proc. of WWW.* DOI: 10.1145/2187836.2187931. 138

Rajarshi Das, Arvind Neelakantan, David Belanger, and Andrew McCallum, (2016). Chains of reasoning over entities, relations, and text using recurrent neural networks. *ArXiv Preprint ArXiv:1607.01426.* DOI: 10.18653/v1/e17-1013. 134

Hal Daume III, (2007). Frustratingly easy domain adaptation. In *ACL*, pages 256–263. 22

Hal Daumé III, (2009). Bayesian multitask learning with latent hierarchies. In *UAI*, pages 135–142. 26

Grégoire Mesnil Yann Dauphin, Xavier Glorot, Salah Rifai, Yoshua Bengio, Ian Goodfellow, Erick Lavoie, Xavier Muller, Guillaume Desjardins, David Warde-Farley, and Pascal Vincent, (2012). Unsupervised and transfer learning challenge: a deep learning approach. In *Proc. of ICML Workshop on Unsupervised and Transfer Learning*, pages 97–110. 56

Marc Peter Deisenroth, Peter Englert, Jochen Peters, and Dieter Fox, (2014). Multi-task policy search for robotics. In *ICRA*, pages 3876–3881. DOI: 10.1109/icra.2014.6907421. 8, 140

Teo de Campos, Bodla Rakesh Babu, and Manik Varma, (2009). Character recognition in natural images. 75
DOI: 10.5220/0001770102730280.

Rocco De Rosa, Thomas Mensink, and Barbara Caputo, (2016). Online open world recognition. *ArXiv:1604.02275 [cs.CV].* 77

Chuong Do and Andrew Y. Ng, (2005). Transfer learning for text classification. In *NIPS*, pages 299–306. DOI: 10.1007/978-3-642-05224-8_3. 23

Jeff Donahue, Yangqing Jia, Oriol Vinyals, Judy Hoffman, Ning Zhang, Eric Tzeng, and Trevor Darrell, (2014). Decaf: A deep convolutional activation feature for generic visual recognition. In *ICML*, pages 647–655. 56

Mark Dredze and Koby Crammer, (2008). Online methods for multi-domain learning and adaptation. In *EMNLP*, pages 689–697. DOI: 10.3115/1613715.1613801. 31

Yan Duan, Marcin Andrychowicz, Bradly Stadie, Jonathan Ho, Jonas Schneider, Ilya Sutskever, Pieter Abbeel, and Wojciech Zaremba. One-shot imitation learning. In *NIPS*, pages 1087–1098, 2017. 34

Vladimir Eidelman, Jordan Boyd-Graber, and Philip Resnik, (2012). Topic models for dynamic translation model adaptation. In *ACL*, pages 115–119. 91

Mathias Eitz, James Hays, and Marc Alexa, (2012). How do humans sketch objects? DOI: 10.1145/2185520.2335395. 75

Salam El Bsat and Matthew E. Taylor, (2017). Scalable multitask policy gradient reinforcement learning. In *AAAI*. 140

Oren Etzioni, Michael Cafarella, Doug Downey, Stanley Kok, Ana-Maria Popescu, Tal Shaked, Stephen Soderland, Daniel S. Weld, and Alexander Yates, (2004). Web-scale information extraction in knowitall: (preliminary results). In *WWW*, pages 100–110. DOI: 10.1145/988672.988687. 111

Theodoros Evgeniou and Massimiliano Pontil, (2004). Regularized multi-task learning. In *KDD*, pages 109–117. DOI: 10.1145/1014052.1014067. 26, 37

Geli Fei and Bing Liu, (2015). Social media text classification under negative covariate shift. In *EMNLP*, pages 2347–2356. DOI: 10.18653/v1/d15-1282. 84

Geli Fei and Bing Liu, (2016). Breaking the closed world assumption in text classification. In *Proc. of NAACL-HLT*, pages 506–514. DOI: 10.18653/v1/n16-1061. 7, 77

Geli Fei, Zhiyuan Chen, and Bing Liu, (2014). Review topic discovery with phrases using the Pólya urn model. In *COLING*, pages 667–676. 91

Geli Fei, Shuai Wang, and Bing Liu, (2016). Learning cumulatively to become more knowledgeable. In *KDD*. DOI: 10.1145/2939672.2939835. 7, 11, 15, 25, 64, 77, 78, 79, 80, 81, 82, 83, 89, 156

Fernando Fernández and Manuela Veloso, (2013). Learning domain structure through probabilistic policy reuse in reinforcement learning. *Progress in Artificial Intelligence*, 2(1), pages 13–27. DOI: 10.1007/s13748-012-0026-6. 8, 140

Chrisantha Fernando, Dylan Banarse, Charles Blundell, Yori Zwols, David Ha, Andrei A. Rusu, Alexander Pritzel, and Daan Wierstra, (2017). Pathnet: Evolution channels gradient descent in super neural networks. *ArXiv Peprint ArXiv:1701.08734*. 58, 72, 74

Chelsea Finn, Pieter Abbeel, and Sergey Levine. Model-agnostic meta-learning for fast adaptation of deep networks. In *ICML*, 2017. 33

Tommaso Furlanello, Jiaping Zhao, Andrew M. Saxe, Laurent Itti, and Bosco S. Tjan, (2016). Active long term memory networks. *ArXiv Preprint ArXiv:1606.02355*. 58

Eli M. Gafni and Dimitri P. Bertsekas, (1984). Two-metric projection methods for constrained optimization. *SIAM Journal on Control and Optimization*, 22(6), pages 936–964. DOI: 10.1137/0322061. 28

Jing Gao, Wei Fan, Jing Jiang, and Jiawei Han, (2008). Knowledge transfer via multiple model local structure mapping. In *KDD*, pages 283–291. DOI: 10.1145/1401890.1401928. 22

Matt Gardner and Tom M. Mitchell, (2015). Efficient and expressive knowledge base completion using subgraph feature extraction. In *EMNLP*. DOI: 10.18653/v1/d15-1173. 133

Jort F. Gemmeke, Daniel P. W. Ellis, Dylan Freedman, Aren Jansen, Wade Lawrence, R. Channing Moore, Manoj Plakal, and Marvin Ritter, (2017). Audio set: An ontology and human-labeled dataset for audio events. In *ICASSP*, pages 776–780, IEEE. DOI: 10.1109/icassp.2017.7952261. 73, 75

Alexander Gepperth and Cem Karaoguz, (2016). A bio-inspired incremental learning architecture for applied perceptual problems. *Cognitive Computation*, 8(5):924–934. DOI: 10.1007/s12559-016-9389-5. 58, 73

Marjan Ghazvininejad, Chris Brockett, Ming-Wei Chang, Bill Dolan, Jianfeng Gao, Wen-tau Yih, and Michel Galley, (2017). A knowledge-grounded neural conversation model. *ArXiv Preprint ArXiv:1702.01932*. 131

Xavier Glorot, Antoine Bordes, and Yoshua Bengio, (2011). Domain adaptation for large-scale sentiment classification: A deep learning approach. In *ICML*, pages 513–520. 24

Pinghua Gong, Jieping Ye, and Changshui Zhang, (2012). Robust multi-task feature learning. In *KDD*, pages 895–903. DOI: 10.1145/2339530.2339672. 27

Ian Goodfellow, (2016). NIPS 2016 tutorial: Generative adversarial networks. *ArXiv Preprint ArXiv:1701.00160*. 57, 58, 70, 71

Ian Goodfellow, Mehdi Mirza, Da Xiao, Aaron Courville, and Yoshua Bengio, (2013a). An empirical investigation of catastrophic forgetting in gradient-based neural networks. *ArXiv Preprint ArXiv:1312.6211*. 59, 72, 74

Ian Goodfellow, Jean Pouget-Abadie, Mehdi Mirza, Bing Xu, David Warde-Farley, Sherjil Ozair, Aaron Courville, and Yoshua Bengio, (2014). Generative adversarial nets. In *NIPS*, pages 2672–2680. 70, 72

Ian Goodfellow, Yoshua Bengio, and Aaron Courville, (2016). *Deep Learning*. MIT Press. http://www.deeplearningbook.org DOI: 10.1038/nature14539. 85

Ian Goodfellow, David Warde-Farley, Mehdi Mirza, Aaron Courville, and Yoshua Bengio, (2013b). Maxout networks. *ArXiv Preprint ArXiv:1302.4389*. 72

Ben Goodrich and Itamar Arel, (2014). Unsupervised neuron selection for mitigating catastrophic forgetting in neural networks. In *Circuits and Systems (MWSCAS), IEEE 57th International Midwest Symposium on*, pages 997–1000. DOI: 10.1109/mwscas.2014.6908585. 58

Gregory Griffin, Alex Holub, and Pietro Perona, (2007). Caltech-256 object category dataset. 75

Erin Grant, Chelsea Finn, Sergey Levine, Trevor Darrell, and Thomas Griffiths. Recasting gradient based meta-learning as hierarchical Bayes. In *arXiv preprint arXiv:1801.08930*, 2018. 34

Thomas L. Griffiths and Mark Steyvers, (2004). Finding scientific topics. *PNAS*, 101 Suppl., pages 5228–5235. DOI: 10.1073/pnas.0307752101. 95

R. He and J. McAuley, (2016). Ups and downs: Modeling the visual evolution of fashion trends with one-class collaborative filtering. In *WWW*. DOI: 10.1145/2872427.2883037. 54

Xu He and Herbert Jaeger, (2018). Overcoming Catastrophic Interference using Conceptor-Aided Backpropagation. In *International Conference on Learning Representations*. DOI: 10.1371/journal.pone.0105619. 58

James J. Heckman, (1979). Sample selection bias as a specification error. *Econometrica: Journal of the Econometric Society*, pages 153–161. DOI: 10.2307/1912352. 47

Gregor Heinrich, (2009). A generic approach to topic models. In *ECML PKDD*, pages 517–532. DOI: 10.1007/978-3-642-04180-8_51. 98

Mark Herbster, Massimiliano Pontil, and Lisa Wainer, (2005). Online learning over graphs. In *ICML*, pages 305–312. DOI: 10.1145/1102351.1102390. 31

Geoffrey E. Hinton, Nitish Srivastava, Alex Krizhevsky, Ilya Sutskever, and Ruslan R. Salakhutdinov, (2012). Improving neural networks by preventing co-adaptation of feature detectors. *ArXiv Preprint ArXiv:1207.0580*. 59, 72

Geoffrey Hinton, Oriol Vinyals, and Jeff Dean, (2015). Distilling the knowledge in a neural network. *ArXiv Preprint ArXiv:1503.02531*. 60, 65

Sepp Hochreiter and Jürgen Schmidhuber, (1997). Long short-term memory. In *Neural computation*, 9(8):1735–1780, 1997. 33

Matthew Hoffman, Francis R. Bach, and David M. Blei, (2010). Online learning for latent Dirichlet allocation. In *NIPS*, pages 856–864. 31

Thomas Hofmann, (1999). Probabilistic latent semantic analysis. In *UAI*, pages 289–296. DOI: 10.1145/312624.312649. 91

Yuening Hu, Jordan Boyd-Graber, and Brianna Satinoff, (2011). Interactive topic modeling. In *ACL*, pages 248–257. DOI: 10.1007/s10994-013-5413-0. 92

Jui-Ting Huang, Jinyu Li, Dong Yu, Li Deng, and Yifan Gong, (2013a). Cross-language knowledge transfer using multilingual deep neural network with shared hidden layers. In *IEEE International Conference on Acoustics, Speech and Signal Processing*, pages 7304–7308. DOI: 10.1109/icassp.2013.6639081. 30

Eric H. Huang, Richard Socher, Christopher D. Manning, and Andrew Y. Ng, (2012). Improving word representations via global context and multiple word prototypes. In *ACL*, pages 873–882. 121

Yan Huang, Wei Wang, Liang Wang, and Tieniu Tan, (2013b). Multi-task deep neural network for multi-label learning. In *IEEE International Conference on Image Processing*, pages 2897–2900. DOI: 10.1109/icip.2013.6738596. 30

Robert A. Hummel and Steven W. Zucker, (1983). On the foundations of relaxation labeling processes. *IEEE Transactions on Pattern Analysis and Machine Intelligence*, (3), pages 267–287. DOI: 10.1016/b978-0-08-051581-6.50058-1. 127

David Isele, Mohammad Rostami, and Eric Eaton, (2016). Using task features for zero-shot knowledge transfer in lifelong learning. In *IJCAI*. 151

Laurent Jacob, Jean-philippe Vert, and Francis R. Bach, (2009). Clustered multi-task learning: A convex formulation. In *NIPS*, pages 745–752. 27, 28

Jagadeesh Jagarlamudi, Hal Daumé III, and Raghavendra Udupa, (2012). Incorporating lexical priors into topic models. In *EACL*, pages 204–213. 92

Kevin Jarrett, Koray Kavukcuoglu, Yann LeCun, et al., (2009). What is the best multi-stage architecture for object recognition? In *Computer Vision, IEEE 12th International Conference on*, pages 2146–2153. DOI: 10.1109/iccv.2009.5459469. 72

Jing Jiang and ChengXiang Zhai, (2007). Instance weighting for domain adaptation in NLP. In *ACL*, pages 264–271.

Jing Jiang, (2008). A literature survey on domain adaptation of statistical classifiers. *Technical Report*. 21

Yaochu Jin and Bernhard Sendhoff, (2006). Alleviating catastrophic forgetting via multi-objective learning. In *IJCNN*, pages 3335–3342. DOI: 10.1109/ijcnn.2006.247332. 10, 21

Nitin Jindal and Bing Liu, (2008). Opinion spam and analysis. In *WSDM*, pages 219–230. DOI: 10.1145/1341531.1341560. 58

Heechul Jung, Jeongwoo Ju, Minju Jung, and Junmo Kim, (2016). Less-forgetting learning in deep neural networks. *ArXiv Preprint ArXiv:1607.00122*. 121

Leslie Pack Kaelbling, Michael L. Littman, and Andrew W. Moore, (1996). Reinforcement learning: A survey. *Journal of Artificial Intelligence Research*, pages 237–285. 57, 74

Nitin Kamra, Umang Gupta, and Yan Liu, (2017). Deep generative dual memory network for continual learning. *ArXiv Preprint ArXiv:1710.10368*. 32, 139

Zhuoliang Kang, Kristen Grauman, and Fei Sha, (2011). Learning with whom to share in multi-task feature learning. In *ICML*, pages 521–528. 58

Christos Kaplanis, Murray Shanahan, and Claudia Clopath, (2018). Continual reinforcement learning with complex synapses. *ArXiv Preprint ArXiv:1802.07239*. 27

Ashish Kapoor and Eric Horvitz, (2009). Principles of lifelong learning for predictive user modeling. In *User Modeling*, pages 37–46. DOI: 10.1007/978-3-540-73078-1_7. 59

Ronald Kemker and Christopher Kanan, (2018). FearNet: Brain-inspired model for incremental learning. In *ICLR*.

Ronald Kemker, Angelina Abitino, Marc McClure, and Christopher Kanan, (2018). Measuring catastrophic forgetting in neural networks. In *AAAI*. 58

Yoon Kim, (2014). Convolutional neural networks for sentence classification. *ArXiv Preprint ArXiv:1408.5882*. DOI: 10.3115/v1/d14-1181. 59, 72, 73, 74, 75

Hajime Kimura, (1995). Reinforcement learning by stochastic hill climbing on discounted reward. In *ICML*, pages 295–303. DOI: 10.1016/b978-1-55860-377-6.50044-x. 85

Diederik P. Kingma and Max Welling, (2013). Auto-encoding variational bayes. *ArXiv Preprint ArXiv:1312.6114*. 141

James Kirkpatrick, Razvan Pascanu, Neil Rabinowitz, Joel Veness, Guillaume Desjardins, Andrei A Rusu, Kieran Milan, John Quan, Tiago Ramalho, Agnieszka Grabska-Barwinska, et al., (2017). Overcoming catastrophic forgetting in neural networks. *Proc. of the National Academy of Sciences*, 114(13):3521–3526. DOI: 10.1073/pnas.1611835114. 72

Jyrki Kivinen, Alexander J. Smola, and Robert C. Williamson, (2004). Online learning with kernels. *IEEE Transactions on Signal Processing*, 52(8), pages 2165–2176. DOI: 10.1109/tsp.2004.830991. 57, 58, 59, 62, 64, 72, 73, 74, 75

Jens Kober and Jan Peters, (2011). Policy search for motor primitives in robotics. *Machine Learning*, 84(1), pages 171–203. DOI: 10.1007/s10994-010-5223-6. 31

George Konidaris and Andrew Barto, (2006). Autonomous shaping: Knowledge transfer in reinforcement learning. In *ICML*, pages 489–496. DOI: 10.1145/1143844.1143906. 148

Alex Krizhevsky and Geoffrey Hinton, (2009). Learning multiple layers of features from tiny images. 140
74

Alex Krizhevsky, Ilya Sutskever, and Geoffrey E. Hinton, (2012). Imagenet classification with deep convolutional neural networks. In *NIPS*, pages 1097–1105. DOI: 10.1145/3065386. 56

Abhishek Kumar, Hal Daum, and Hal Daume Iii, (2012). Learning task grouping and overlap in multi-task learning. In *ICML*, pages 1383–1390. 7, 27, 28, 36, 41

Dharshan Kumaran, Demis Hassabis, and James L. McClelland, (2016). What learning systems do intelligent agents need? Complementary learning systems theory updated. *Trends in Cognitive Sciences*, 20(7):512–534. DOI: 10.1016/j.tics.2016.05.004. 58

John Lafferty, Andrew McCallum, and Fernando C. N. Pereira, (2001). Conditional random fields: Probabilistic models for segmenting and labeling sequence data. In *ICML*, pages 282–289. 123

Ni Lao, Tom Mitchell, and William W. Cohen, (2011). Random walk inference and learning in a large scale knowledge base. In *EMNLP*. 132, 133

Ni Lao, Einat Minkov, and William W. Cohen, (2015). Learning relational features with backward random walks. In *ACL*. DOI: 10.3115/v1/p15-1065. 132

Neil D. Lawrence and John C. Platt, (2004). Learning to learn with the informative vector machine. In *ICML*. DOI: 10.1145/1015330.1015382. 22

Alessandro Lazaric and Mohammad Ghavamzadeh, (2010). Bayesian multi-task reinforcement learning. In *ICML*, pages 599–606. 10, 32

Phong Le, Marc Dymetman, and Jean-Michel Renders, (2016). LSTM-based mixture-of-experts for knowledge-aware dialogues. *ArXiv Preprint ArXiv:1605.01652*. DOI: 10.18653/v1/w16-1611. 131

Yann LeCun, John S. Denker, and Sara A. Solla, (1990). Optimal brain damage. In *NIPS*, pages 598–605. 61

Yann LeCun, Léon Bottou, Yoshua Bengio, and Patrick Haffner, (1998). Gradient-based learning applied to document recognition. *Proc. of the IEEE*, 86(11):2278–2324. DOI: 10.1109/5.726791. 64, 65, 72, 73, 74

Jeongtae Lee, Jaehong Yun, Sungju Hwang, and Eunho Yang, (2017a). Lifelong learning with dynamically expandable networks. *ArXiv Preprint ArXiv:1708.01547*. 58

Su-In Lee, Vassil Chatalbashev, David Vickrey, and Daphne Koller, (2007). Learning a meta-level prior for feature relevance from multiple related tasks. In *ICML*, pages 489–496. DOI: 10.1145/1273496.1273558. 26

Sang-Woo Lee, Jin-Hwa Kim, Jaehyun Jun, Jung-Woo Ha, and Byoung-Tak Zhang, (2017b). Overcoming catastrophic forgetting by incremental moment matching. In *Advances in Neural Information Processing Systems*, pages 4655–4665. 58

Zhizhong Li and Derek Hoiem, (2016). Learning without forgetting. In *European Conference on Computer Vision*, pages 614–629, Springer. DOI: 10.1007/978-3-319-46493-0_37. 56, 57, 58, 59, 60, 74, 75

Hui Li, Xuejun Liao, and Lawrence Carin, (2009). Multi-task reinforcement learning in partially observable stochastic environments. *The Journal of Machine Learning Research*, 10, pages 1131–1186. 26, 32

Jiwei Li, Will Monroe, and Dan Jurafsky, (2017a). Data distillation for controlling specificity in dialogue generation. *ArXiv Preprint ArXiv:1702.06703*. 131

Jiwei Li, Will Monroe, Tianlin Shi, Alan Ritter, and Dan Jurafsky, (2017b). Adversarial learning for neural dialogue generation. *ArXiv Preprint ArXiv:1701.06547*. DOI: 10.18653/v1/d17-1230. 131

Zhenguo Li, Fengwei Zhou, Fei Chen, and Hang Li. Meta-sgd: Learning to learn quickly for few shot learning. In *arXiv preprint arXiv:1707.09835*, 2017c. 34

Xuejun Liao, Ya Xue, and Lawrence Carin, (2005). Logistic regression with an auxiliary data source. In *ICML*, pages 505–512. DOI: 10.1145/1102351.1102415. 21

Zachary C. Lipton, Jianfeng Gao, Lihong Li, Jianshu Chen, and Li Deng, (2016). Combating reinforcement learning's sisyphean curse with intrinsic fear. *ArXiv Preprint ArXiv:1611.01211*. 59, 75

Bing Liu, (2007). *Web Data Mining: exploring hyperlinks, contents, and usage data*. Springer. DOI: 10.1007/978-3-642-19460-3. 102, 119

Bing Liu, (2012). Sentiment Analysis and Opinion Mining. *Synthesis Lectures on Human Language Technologies*, Morgan & Claypool Publishers. xvii, 3, 35, 91, 92, 117

Bing Liu, (2015). *Sentiment Analysis: Mining Opinions, Sentiments, and Emotions*. Cambridge University Press. xvii, 3, 92

Qian Liu, Zhiqiang Gao, Bing Liu, and Yuanlin Zhang, (2015a). Automated rule selection for aspect extraction in opinion mining. In *IJCAI*, pages 1291–1297. 129

Bing Liu, Wynne Hsu, and Yiming Ma, (1999). Mining association rules with multiple minimum supports. In *KDD*, pages 337–341, ACM. DOI: 10.1145/312129.312274. 102

Bing Liu, Wee Sun Lee, Philip S. Yu, and Xiaoli Li, (2002). Partially supervised classification
of text documents. In *ICML*, pages 387–394. 116

Qian Liu, Bing Liu, Yuanlin Zhang, Doo Soon Kim, and Zhiqiang Gao, (2016). Improving
opinion aspect extraction using semantic similarity and aspect associations. In *AAAI*. 7, 14,
15, 16, 111, 117, 118, 119, 120, 121, 129

Xiaodong Liu, Jianfeng Gao, Xiaodong He, Li Deng, Kevin Duh, and Ye-Yi Wang, (2015b).
Representation learning using multi-task deep neural networks for semantic classification and
information retrieval. In *NAACL*. DOI: 10.3115/v1/n15-1092. 4, 29, 30

Vincenzo Lomonaco and Davide Maltoni, (2017). CORe50: A new dataset and benchmark for
continuous object recognition. *ArXiv Preprint ArXiv:1705.03550*. 75

David Lopez-Paz et al., (2017). Gradient episodic memory for continual learning. In *Advances
in Neural Information Processing Systems*, pages 6470–6479. 58, 74

Ryan Lowe, Nissan Pow, Iulian Serban, and Joelle Pineau, (2015). The ubuntu dialogue corpus:
A large dataset for research in unstructured multi-turn dialogue systems. *ArXiv Preprint
ArXiv:1506.08909*. DOI: 10.18653/v1/w15-4640. 131, 138

Justin Ma, Lawrence K. Saul, Stefan Savage, and Geoffrey M. Voelker, (2009). Identifying
suspicious URLs: An application of large-scale online learning. In *ICML*, pages 681–688.
DOI: 10.1145/1553374.1553462. 31

Hosam Mahmoud, (2008). *Polya Urn Models*. Chapman & Hall/CRC Texts in Statistical
Science. DOI: 10.1201/9781420059847. 97, 98

Julien Mairal, Francis Bach, Jean Ponce, and Guillermo Sapiro, (2009). Online dictionary learning for sparse coding. In *ICML*, pages 689–696. DOI: 10.1145/1553374.1553463. 31

Julien Mairal, Francis Bach, Jean Ponce, and Guillermo Sapiro, (2010). Online learning for
matrix factorization and sparse coding. *The Journal of Machine Learning Research*, 11, pages 19–
60. 31

S. Maji, J. Kannala, E. Rahtu, M. Blaschko, and A. Vedaldi, (2013). Fine-grained visual classification of aircraft. *Technical Report*. 75

Arun Mallya and Svetlana Lazebnik, (2017). PackNet: Adding multiple tasks to a single network by iterative pruning. *ArXiv Preprint ArXiv:1711.05769*. 58

Daniel J. Mankowitz, Augustin Žídek, André Barreto, Dan Horgan, Matteo Hessel, John Quan, Junhyuk Oh, Hado van Hasselt, David Silver, and Tom Schaul, (2018). Unicorn: Continual learning with a universal, off-policy agent. *ArXiv Preprint ArXiv:1802.08294*. 59, 75

Christopher D. Manning, Prabhakar Raghavan, Hinrich Schütze, et al., (2008). *Introduction to Information Retrieval*, vol. 1. Cambridge University Press, Cambridge. DOI: 10.1017/cbo9780511809071. 84

Nicolas Y. Masse, Gregory D. Grant, and David J. Freedman, (2018). Alleviating catastrophic forgetting using context-dependent gating and synaptic stabilization. *ArXiv Preprint ArXiv:1802.01569*. 58

Andreas Maurer, Massimiliano Pontil, and Bernardino Romera-Paredes, (2013). Sparse coding for multitask and transfer learning. In *ICML*, pages 343–351. 27

Sahisnu Mazumder and Bing Liu, (2017). Context-aware path ranking for knowledge base completion. In *IJCAI*. DOI: 10.24963/ijcai.2017/166. 132, 134

Sahisnu Mazumder, Nianzu Ma, and Bing Liu, (2018). Towards a continuous knowledge learning engine for chatbots. In *ArXiv:1802.06024 [cs.CL]*. 7, 131, 132, 133, 134, 135, 137, 138

Andrew McCallum and Kamal Nigam, (1998). A comparison of event models for Naive Bayes text classification. In *AAAI Workshop Learning for Text Categorization*. 47

James L. McClelland, Bruce L. McNaughton, and Randall C. O'reilly, (1995). Why there are complementary learning systems in the hippocampus and neocortex: Insights from the successes and failures of connectionist models of learning and memory. *Psychological Review*, 102(3):419. DOI: 10.1037//0033-295x.102.3.419. 58

Michael McCloskey and Neal J. Cohen, (1989). Catastrophic interference in connectionist networks: The sequential learning problem. In *Psychology of Learning and Motivation*, vol. 24, pages 109–165, Elsevier. DOI: 10.1016/s0079-7421(08)60536-8. 7, 55

Neville Mehta, Sriraam Natarajan, Prasad Tadepalli, and Alan Fern, (2008). Transfer in variable-reward hierarchical reinforcement learning. *Machine Learning*, 73(3), pages 289–312. DOI: 10.1007/s10994-008-5061-y. 32

Thomas Mensink, Jakob Verbeek, Florent Perronnin, and Gabriela Csurka, (2013). Distance based image classification: Generalizing to new classes at near-zero cost. *IEEE Transactions Pattern Analysis and Machine Intelligence*, 35(11):2624—2637. DOI: 10.1109/tpami.2013.83. 79

Tomas Mikolov, Kai Chen, Greg Corrado, and Jeffrey Dean, (2013a). Efficient estimation of word representations in vector space. *ArXiv*. 51

Tomas Mikolov, Ilya Sutskever, Kai Chen, Greg S. Corrado, and Jeff Dean, (2013). Distributed representations of words and phrases and their compositionality. In *NIPS*, pages 3111–3119.

Tomas Mikolov, Ilya Sutskever, Kai Chen, Greg S. Corrado, and Jeff Dean, (2013b). Distributed representations of words and phrases and their compositionality. In *Advances in Neural Information Processing Systems 26*, pages 3111–3119, Curran Associates, Inc. 51, 120

George A. Miller, (1995). WordNet: A lexical database for English. *Communications on ACM*, 38(11), pages 39–41. DOI: 10.1145/219717.219748. 92

David Mimno, Hanna M. Wallach, Edmund Talley, Miriam Leenders, and Andrew McCallum, (2011). Optimizing semantic coherence in topic models. In *EMNLP*, pages 262–272. 98, 99

Nikhil Mishra, Mostafa Rohaninejad, Xi Chen, and Pieter Abbeel. A simple neural attentive meta-learner. In *ICLR*, 2018. 34

T. Mitchell, W. Cohen, E. Hruschka, P. Talukdar, J. Betteridge, A. Carlson, B. Dalvi, M. Gardner, B. Kisiel, J. Krishnamurthy, N. Lao, K. Mazaitis, T. Mohamed, N. Nakashole, E. Platanios, A. Ritter, M. Samadi, B. Settles, R. Wang, D. Wijaya, A. Gupta, X. Chen, A. Saparov, M. Greaves, and J. Welling, (2015). Never-ending learning. In *AAAI*. DOI: 10.1145/3191513. 8, 16, 111, 115, 116, 154, 155

Andriy Mnih and Geoffrey Hinton, (2007). Three new graphical models for statistical language modelling. In *ICML*. DOI: 10.1145/1273496.1273577. 51

Volodymyr Mnih, Koray Kavukcuoglu, David Silver, Alex Graves, Ioannis Antonoglou, Daan Wierstra, and Martin Riedmiller, (2013). Playing atari with deep reinforcement learning. *ArXiv Preprint ArXiv:1312.5602.* 75

Joseph Modayil, Adam White, and Richard S. Sutton, (2014). Multi-timescale nexting in a reinforcement learning robot. *Adaptive Behavior*, 22(2), pages 146–160. DOI: 10.1177/1059712313511648. 33

Arjun Mukherjee and Bing Liu, (2012). Aspect extraction through semi-supervised modeling. In *ACL*, pages 339–348. 91, 92

Stefan Munder and Dariu M. Gavrila, (2006). An experimental study on pedestrian classification. *IEEE Transactions on Pattern Analysis and Machine Intelligence*, 28(11):1863–1868. DOI: 10.1109/tpami.2006.217. 75

Tsendsuren Munkhdalai and Hong Yu. Meta networks. *arXiv preprint arXiv:1703.00837*, 2017. 33

Arvind Neelakantan, Benjamin Roth, and Andrew McCallum, (2015). Compositional vector space models for knowledge base completion. *ArXiv Preprint ArXiv:1504.06662.* DOI: 10.3115/v1/p15-1016. 134

Yuval Netzer, Tao Wang, Adam Coates, Alessandro Bissacco, Bo Wu, and Andrew Y. Ng, (2011). Reading digits in natural images with unsupervised feature learning. In *NIPS Workshop on Deep Learning and Unsupervised Feature Learning*, page 5. 74

Cuong V. Nguyen, Yingzhen Li, Thang D. Bui, and Richard E. Turner, (2017). Variational continual learning. *ArXiv Preprint ArXiv:1710.10628*. 58

Maximilian Nickel, Kevin Murphy, Volker Tresp, and Evgeniy Gabrilovich, (2015). A review of relational machine learning for knowledge graphs. *ArXiv Preprint ArXiv:1503.00759*. DOI: 10.1109/jproc.2015.2483592. 132

Maria-Elena Nilsback and Andrew Zisserman, (2008). Automated flower classification over a large number of classes. In *6th Indian Conference on Computer Vision, Graphics and Image Processing*, pages 722–729, IEEE. DOI: 10.1109/icvgip.2008.47. 75

Sinno Jialin Pan and Qiang Yang, (2010). A survey on transfer learning. *IEEE Transactions on Knowledge and Data Engineering*, 22(10), pages 1345–1359. DOI: 10.1109/tkde.2009.191. 10, 21

Sinno Jialin Pan, Xiaochuan Ni, Jian-Tao Sun, Qiang Yang, and Zheng Chen, (2010). Cross-domain sentiment classification via spectral feature alignment. In *WWW*, pages 751–760. DOI: 10.1145/1772690.1772767. 22

German I. Parisi, Ronald Kemker, Jose L. Part, Christopher Kanan, and Stefan Wermter, (2018a). Continual lifelong learning with neural networks: A review. *ArXiv Preprint ArXiv:1802.07569*. 7, 57

German I. Parisi, Jun Tani, Cornelius Weber, and Stefan Wermter, (2017). Lifelong learning of human actions with deep neural network self-organization. *Neural Networks*, 96:137–149. DOI: 10.1016/j.neunet.2017.09.001. 59

German I. Parisi, Jun Tani, Cornelius Weber, and Stefan Wermter, (2018b). Lifelong learning of spatiotemporal representations with dual-memory recurrent self-organization. *ArXiv Preprint ArXiv:1805.10966*. 58, 75

Jeffrey Pennington, Richard Socher, and Christopher D. Manning, (2014). Glove: Global vectors for word representation. In *EMNLP*, pages 1532–1543. DOI: 10.3115/v1/d14-1162. 51, 121

Anastasia Pentina and Christoph H. Lampert, (2014). A PAC-Bayesian bound for lifelong learning. In *ICML*, pages 991–999. 7, 36

Jan Peters and Stefan Schaal, (2006). Policy gradient methods for robotics. In *IROS*, pages 2219–2225. DOI: 10.1109/iros.2006.282564. 147

Jan Peters and J. Andrew Bagnell, (2011). Policy gradient methods. In *Encyclopedia of Machine Learning*, pages 774–776, Springer. DOI: 10.1007/978-1-4899-7502-7_646-1. 147

James Petterson, Alex Smola, Tibério Caetano, Wray Buntine, and Shravan Narayanamurthy, (2010). Word features for latent Dirichlet allocation. In *NIPS*, pages 1921–1929. 92

John Platt et al., (1999). Probabilistic outputs for support vector machines and comparisons to regularized likelihood methods. *Advances in Large Margin Classifiers*, 10(3), pages 61–74. DOI: 10.1016/j.knosys.2012.04.006. 82

Robi Polikar, Lalita Upda, Satish S. Upda, and Vasant Honavar, (2001). Learn++: An incremental learning algorithm for supervised neural networks. *IEEE Transactions on Systems, Man, and Cybernetics, Part C (Applications and Reviews)*, 31(4):497–508. DOI: 10.1109/5326.983933. 58

Dean A. Pomerleau, (2012). *Neural Network Perception for Mobile Robot Guidance*, vol. 239. Springer Science and Business Media. DOI: 10.1007/978-1-4615-3192-0. 53

Guang Qiu, Bing Liu, Jiajun Bu, and Chun Chen, (2011). Opinion word expansion and target extraction through double propagation. *Computational Linguistics*, 37(1), pages 9–27. DOI: 10.1162/coli_a_00034. 118

Ariadna Quattoni and Antonio Torralba, (2009). Recognizing indoor scenes. In *CVPR*, pages 413–420, IEEE. DOI: 10.1109/cvpr.2009.5206537. 75

J. Ross Quinlan and R. Mike Cameron-Jones, (1993). FOIL: A midterm report. In *ECML*, pages 3–20. DOI: 10.1007/3-540-56602-3_124. 116, 117

Alec Radford, Luke Metz, and Soumith Chintala, (2015). Unsupervised representation learning with deep convolutional generative adversarial networks. *ArXiv Preprint ArXiv:1511.06434*. 70

Rajat Raina, Alexis Battle, Honglak Lee, Benjamin Packer, and Andrew Y. Ng, (2007). Self-taught learning: Transfer learning from unlabeled data. In *ICML*, pages 759–766. DOI: 10.1145/1273496.1273592. 156

Steve Ramirez, Xu Liu, Pei-Ann Lin, Junghyup Suh, Michele Pignatelli, Roger L. Redondo, Tomás J. Ryan, and Susumu Tonegawa, (2013). Creating a false memory in the hippocampus. *Science*, 341(6144):387–391. 70
DOI: 10.1126/science.1239073.

Amal Rannen Ep Triki, Rahaf Aljundi, Matthew Blaschko, and Tinne Tuytelaars, (2017). Encoder based lifelong learning. In *ICCV 2017*, pages 1320–1328. DOI: 10.1109/iccv.2017.148. 57, 70, 74

Sachin Ravi and Hugo Larochelle. Optimization as a model for few-shot learning. In *ICLR*, 2017. 34

Sylvestre-Alvise Rebuffi, Alexander Kolesnikov, and Christoph H. Lampert, (2017). iCaRL: Incremental classifier and representation learning. In *CVPR*, pages 5533–5542. DOI: 10.1109/cvpr.2017.587. 57, 64, 65, 74, 75

Fiona M. Richardson and Michael S. C. Thomas, (2008). Critical periods and catastrophic interference effects in the development of self-organizing feature maps. *Developmental Science*, 11(3):371–389. DOI: 10.1111/j.1467-7687.2008.00682.x. 56

Leonardo Rigutini, Marco Maggini, and Bing Liu, (2005). An EM based training algorithm for cross-language text categorization. In *Proc. of the IEEE/WIC/ACM International Conference on Web Intelligence*, pages 529–535. DOI: 10.1109/wi.2005.29. 23, 24

Mark Bishop Ring, (1994). Continual learning in reinforcement environments. Ph.D. thesis, University of Texas at Austin Austin, TX. 59

Mark B. Ring, (1998). CHILD: A first step towards continual learning. In *Learning to Learn*, pages 261–292. DOI: 10.1007/978-1-4615-5529-2_11. 8, 140

Anthony Robins, (1995). Catastrophic forgetting, rehearsal and pseudorehearsal. *Connection Science*, 7(2):123–146. DOI: 10.1080/09540099550039318. 58, 71

Amir Rosenfeld and John K. Tsotsos, (2017). Incremental Learning Through Deep Adaptation. *ArXiv Preprint ArXiv:1705.04228*. 57, 74

David E. Rumelhart, Geoffrey E. Hinton, and Ronald J. Williams, (1985). Learning internal representations by error propagation. *Technical Report*, DTIC Document. DOI: 10.21236/ada164453. 39

Olga Russakovsky, Jia Deng, Hao Su, Jonathan Krause, Sanjeev Satheesh, Sean Ma, Zhiheng Huang, Andrej Karpathy, Aditya Khosla, Michael Bernstein, et al., (2015). Imagenet large scale visual recognition challenge. *International Journal of Computer Vision*, 115(3):211–252. DOI: 10.1007/s11263-015-0816-y. 56, 75

Andrei A. Rusu, Sergio Gomez Colmenarejo, Caglar Gulcehre, Guillaume Desjardins, James Kirkpatrick, Razvan Pascanu, Volodymyr Mnih, Koray Kavukcuoglu, and Raia Hadsell, (2015). Policy distillation. *ArXiv Preprint ArXiv:1511.06295*. 61

Andrei A. Rusu, Neil C. Rabinowitz, Guillaume Desjardins, Hubert Soyer, James Kirkpatrick, Koray Kavukcuoglu, Razvan Pascanu, and Raia Hadsell, (2016). Progressive neural networks. *ArXiv Preprint ArXiv:1606.04671*. 57, 59, 61, 75

Paul Ruvolo and Eric Eaton, (2013a). Active task selection for lifelong machine learning. In *AAAI*, pages 862–868. 7, 40, 45, 46

Paul Ruvolo and Eric Eaton, (2013b). ELLA: An efficient lifelong learning algorithm. In *ICML*, pages 507–515. 7, 8, 14, 15, 16, 27, 36, 37, 40, 41, 42, 44, 53, 140, 146, 150

Adam Santoro, Sergey Bartunov, Matthew Botvinick, Daan Wierstra, and Timothy Lillicrap. Meta-learning with memory-augmented neural networks. In *ICML*, pages 1842–1850, 2016. 33

Walter J. Scheirer, Anderson de Rezende Rocha, Archana Sapkota, and Terrance E. Boult, (2013). Toward open set recognition. *Pattern Analysis and Machine Intelligence, IEEE Transactions on*, 35(7), pages 1757–1772. DOI: 10.1109/tpami.2012.256. 82, 83

Juergen Schmidhuber, (2018). One big net for everything. *ArXiv:1802.08864 [cs.AI]*, pages 1–17. 58

Mark Schmidt, Glenn Fung, and Rómer Rosales, (2007). Fast optimization methods for L1 regularization: A comparative study and two new approaches. In *ECML*, pages 286–297. DOI: 10.1007/978-3-540-74958-5_28. 28

Anton Schwaighofer, Volker Tresp, and Kai Yu, (2004). Learning Gaussian process kernels via hierarchical Bayes. In *NIPS*, pages 1209–1216. 22

Ari Seff, Alex Beatson, Daniel Suo, and Han Liu, (2017). Continual learning in generative adversarial nets. *ArXiv Preprint ArXiv:1705.08395*. 58, 74

Michael L. Seltzer and Jasha Droppo, (2013). Multi-task learning in deep neural networks for improved phoneme recognition. In *IEEE International Conference on Acoustics, Speech and Signal Processing*, pages 6965–6969. DOI: 10.1109/icassp.2013.6639012. 30

Iulian Vlad Serban, Ryan Lowe, Peter Henderson, Laurent Charlin, and Joelle Pineau, (2015). A survey of available corpora for building data-driven dialogue systems. *ArXiv Preprint ArXiv:1512.05742*. 131

Joan Serrà, Dídac Surís, Marius Miron, and Alexandros Karatzoglou, (2018). Overcoming catastrophic forgetting with hard attention to the task. *ArXiv Preprint ArXiv:1801.01423*. 58

Nicholas Shackel, (2007). Bertrand's paradox and the principle of indifference. *Philosophy of Science*, 74(2), pages 150–175. DOI: 10.1086/519028. 82

Donald Shepard, (1968). A two-dimensional interpolation function for irregularly-spaced data. In *Proc. of the 23rd ACM National Conference*, pages 517–524. DOI: 10.1145/800186.810616. 37

Hidetoshi Shimodaira, (2000). Improving predictive inference under covariate shift by weighting the log-likelihood function. *Journal of Statistical Planning and Inference*, 90(2), pages 227–244. DOI: 10.1016/s0378-3758(00)00115-4. 47

Hanul Shin, Jung Kwon Lee, Jaehong Kim, and Jiwon Kim, (2017). Continual learning with deep generative replay. In *NIPS*, pages 2994–3003. 57, 58, 70, 71, 74

Lei Shu, Hu Xu, and Bing Liu, (2017a). Doc: Deep open classification of text documents. In *EMNLP*. DOI: 10.18653/v1/d17-1314. 7, 79, 85, 86

Lei Shu, Bing Liu, Hu Xu, and Annice Kim, (2016). Lifelong-RL: Lifelong relaxation labeling for separating entities and aspects in opinion targets using lifelong graph labeling. In *EMNLP*. 7, 14, 15, 16, 127, 129, 154

Lei Shu, Hu Xu, and Bing Liu, (2018). Unseen class discovery in open-world classification. In *ArXiv:1801.05609 [cs.LG]*. 88, 89

Lei Shu, Hu Xu, and Bing Liu, (2017b). Lifelong learning CRF for supervised aspect extraction. In *Proc. of Annual Meeting of the Association for Computational Linguistics (ACL, Short Paper)*. DOI: 10.18653/v1/p17-2023. 7, 14, 25, 111, 123, 129

Daniel L. Silver and Robert E. Mercer, (1996). The parallel transfer of task knowledge using dynamic learning rates based on a measure of relatedness. *Connection Science*, 8(2), pages 277–294. DOI: 10.1007/978-1-4615-5529-2_9. 7, 36

Daniel L. Silver and Robert E. Mercer, (2002). The task rehearsal method of life-long learning: Overcoming impoverished data. In *Proc. of the 15th Conference of the Canadian Society for Computational Studies of Intelligence on Advances in Artificial Intelligence*, pages 90–101. DOI: 10.1007/3-540-47922-8_8. 7, 15, 36, 39

Daniel L. Silver and Ryan Poirier, (2004). Sequential consolidation of learned task knowledge. In *Conference of the Canadian Society for Computational Studies of Intelligence*, pages 217–232. DOI: 10.1007/978-3-540-24840-8_16. 39

Daniel L. Silver and Ryan Poirier, (2007). Context-sensitive MTL networks for machine lifelong learning. In *FLAIRS Conference*, pages 628–633. 39

Daniel L. Silver, Qiang Yang, and Lianghao Li, (2013). Lifelong machine learning systems: Beyond learning algorithms. In *AAAI Spring Symposium: Lifelong Machine Learning*, pages 49–55. 8

Daniel L. Silver, Geoffrey Mason, and Lubna Eljabu, (2015). Consolidation using sweep task rehearsal: Overcoming the stability-plasticity problem. In *Advances in Artificial Intelligence*, vol. 9091, pages 307–322. DOI: 10.1007/978-3-319-18356-5_27. 7, 15, 36

David Silver, Aja Huang, Chris J. Maddison, Arthur Guez, Laurent Sifre, George van den Driessche, Julian Schrittwieser, Ioannis Antonoglou, Veda Panneershelvam, Marc Lanctot, Sander Dieleman, Dominik Grewe, John Nham, Nal Kalchbrenner, Ilya Sutskever, Timothy Lillicrap, Madeleine Leach, Koray Kavukcuoglu, Thore Graepel, and Demis Hassabis, (2016). Mastering the game of Go with deep neural networks and tree search. *Nature*, 529(7587), pages 484–489. DOI: 10.1038/nature16961. 139, 152

David Silver, Julian Schrittwieser, Karen Simonyan, Ioannis Antonoglou, Aja Huang, Arthur Guez, Thomas Hubert, Lucas Baker, Matthew Lai, Adrian Bolton, et al., (2017). Mastering the game of Go without human knowledge. *Nature*, 550(7676):354. DOI: 10.1038/nature24270. 139, 152

Patrice Simard, Bernard Victorri, Yann LeCun, and John Denker, (1992). Tangent prop-a formalism for specifying selected invariances in an adaptive network. In *NIPS*, pages 895–903. 40

Rupesh K. Srivastava, Jonathan Masci, Sohrob Kazerounian, Faustino Gomez, and Jürgen Schmidhuber, (2013). Compete to compute. In *NIPS*, pages 2310–2318. 72

Johannes Stallkamp, Marc Schlipsing, Jan Salmen, and Christian Igel, (2012). Man vs. computer: Benchmarking machine learning algorithms for traffic sign recognition. *Neural Networks*, 32:323–332. DOI: 10.1016/j.neunet.2012.02.016. 75

Robert Stickgold and Matthew P. Walker, (2007). Sleep-dependent memory consolidation and reconsolidation. *Sleep Medicine*, 8(4):331–343. DOI: 10.1016/j.sleep.2007.03.011. 70

Malcolm Strens, (2000). A Bayesian framework for reinforcement learning. In *ICML*, pages 943–950.

Fabian M. Suchanek, Gjergji Kasneci, and Gerhard Weikum, (2007). Yago: A core of semantic knowledge. In *WWW*, pages 697–706. DOI: 10.1145/1242572.1242667. 144

Masashi Sugiyama, Shinichi Nakajima, Hisashi Kashima, Paul V. Buenau, and Motoaki Kawanabe, (2008). Direct importance estimation with model selection and its application to covariate shift adaptation. In *NIPS*, pages 1433–1440. 111

Richard S. Sutton and Andrew G. Barto, (1998). *Reinforcement Learning: An Introduction*. MIT press. DOI: 10.1109/tnn.1998.712192. 21

Richard S. Sutton, David A. McAllester, Satinder P. Singh, Yishay Mansour, et al., (2000). Policy gradient methods for reinforcement learning with function approximation. In *NIPS*, pages 1057–1063. 32, 139, 144

Richard S. Sutton, Joseph Modayil, Michael Delp, Thomas Degris, Patrick M. Pilarski, Adam White, and Doina Precup, (2011). Horde: A scalable real-time architecture for learning knowledge from unsupervised sensorimotor interaction. In *The 10th International Conference on Autonomous Agents and Multiagent Systems*, vol. 2, pages 761–768. 146, 147

Csaba Szepesvári, (2010). Algorithms for reinforcement learning. *Synthesis Lectures on Artificial Intelligence and Machine Learning*, 4(1), pages 1–103. DOI: 10.2200/s00268ed1v01y201005aim009. 32

Fumihide Tanaka and Masayuki Yamamura, (1997). An approach to lifelong reinforcement learning through multiple environments. In *6th European Workshop on Learning Robots*, pages 93–99. 32

Matthew E. Taylor and Peter Stone, (2007). Cross-domain transfer for reinforcement learning. In *ICML*, pages 879–886. DOI: 10.1145/1273496.1273607. 8, 15, 139, 140, 141, 152

Matthew E. Taylor and Peter Stone, (2009). Transfer learning for reinforcement learning domains: A survey. *The Journal of Machine Learning Research*, 10, pages 1633–1685. DOI: 10.1007/978-3-642-01882-4. 32

Matthew E. Taylor, Peter Stone, and Yaxin Liu, (2007). Transfer learning via inter-task mappings for temporal difference learning. *The Journal of Machine Learning Research*, 8, pages 2125–2167.

Matthew E. Taylor, Nicholas K. Jong, and Peter Stone, (2008). Transferring instances for model-based reinforcement learning. In *ECML PKDD*, pages 488–505. DOI: 10.1007/978-3-540-87481-2_32. 10, 21, 32

Chen Tessler, Shahar Givony, Tom Zahavy, Daniel J. Mankowitz, and Shie Mannor, (2017). A deep hierarchical approach to lifelong learning in minecraft. In *AAAI*, vol. 3, page 6. 151

William R. Thompson, (1933). On the likelihood that one unknown probability exceeds another in view of the evidence of two samples. *Biometrika*, 25(3/4), pages 285–294. DOI: 10.2307/2332286. 32

Sebastian Thrun, (1996a). *Explanation-based Neural Network Learning: A Lifelong Learning Approach*. Kluwer Academic Publishers. DOI: 10.1017/s0269888999211034. 140, 152

Sebastian Thrun, (1996b). Is learning the n-th thing any easier than learning the first? In *NIPS*, pages 640–646. 144

Sebastian Thrun and Tom M. Mitchell, (1995). Lifelong Robot Learning. In: L. Steels, editors, *The Biology and Technology of Intelligent Autonomous Agents*, vol 144. Springer. DOI: 10.1007/978-3-642-79629-6_7. 39

Sebastian Thrun. Lifelong learning algorithms. In S. Thrun and L. Pratt, editors, *Learning To Learn*, pages 181–209. Kluwer Academic Publishers, 1998. 6, 9, 15, 36, 37, 38, 40, 53

Geoffrey G. Towell and Jude W. Shavlik, (1994). Knowledge-based artificial neural networks. *Artificial Intelligence*, 70(1-2), pages 119–165. DOI: 10.1016/0004-3702(94)90105-8. 6, 8, 140

Joseph Turian, Lev Ratinov, and Yoshua Bengio, (2010). Word representations: A simple and general method for semi-supervised learning. In *ACL*, pages 384–394. 33

Rasul Tutunov, Julia El-Zini, Haitham Bou-Ammar, and Ali Jadbabaie, (2017). Distributed lifelong reinforcement learning with sub-linear regret. In *Decision and Control (CDC), IEEE 56th Annual Conference on*, pages 2254–2259. DOI: 10.1109/cdc.2017.8263978.

Michel F. Valstar, Bihan Jiang, Marc Mehu, Maja Pantic, and Klaus Scherer, (2011). The first facial expression recognition and analysis challenge. In *IEEE International Conference on Automatic Face and Gesture Recognition*, pages 921–926. DOI: 10.1109/fg.2011.5771374. 51, 120

Roby Velez and Jeff Clune, (2017). Diffusion-based neuromodulation can eliminate catastrophic forgetting in simple neural networks. *PloS One*, 12(11):e0187736. DOI: 10.1371/journal.pone.0187736. 140

Ragav Venkatesan, Hemanth Venkateswara, Sethuraman Panchanathan, and Baoxin Li, (2017). A strategy for an uncompromising incremental learner. *ArXiv Preprint ArXiv:1705.00744.* 53

Ricardo Vilalta and Youssef Drissi. A perspective view and survey of meta-learning. In *Artificial Intelligence Review*, 18(2):77–95, 2002. 58

Pascal Vincent, Hugo Larochelle, Yoshua Bengio, and Pierre-Antoine Manzagol, (2008). Extracting and composing robust features with denoising autoencoders. In *ICML*, pages 1096–1103. DOI: 10.1145/1390156.1390294. 58, 74

Oriol Vinyals and Quoc Le, (2015). A neural conversational model. *ArXiv Preprint ArXiv:1506.05869.* 33

Alexander Waibel, Toshiyuki Hanazawa, Geofrey Hinton, Kiyohiro Shikano, and Kevin J. Lang, (1989). Phoneme recognition using time-delay neural networks. *IEEE Transactions on Acoustics, Speech, and Signal Processing*, pages 328–339. DOI: 10.1016/b978-0-08-051584-7.50037-1. 24

Sida I Wang, Percy Liang, and Christopher D Manning. Learning language games through interaction. In *arXiv preprint arXiv:1606.02447*, 2016. 131

Chang Wang and Sridhar Mahadevan, (2008). Manifold alignment using procrustes analysis. In *ICML*, pages 1120–1127. DOI: 10.1145/1390156.1390297. 30

Chang Wang and Sridhar Mahadevan, (2009). Manifold alignment without correspondence. In *IJCAI*, pages 1273–1278. 34

Richard C. Wang and William W. Cohen, (2009). Character-level analysis of semi-structured documents for set expansion. In *EMNLP*, pages 1503–1512. DOI: 10.3115/1699648.1699697. 22

Tao Wang, Daniel Lizotte, Michael Bowling, and Dale Schuurmans, (2005). Bayesian sparse sampling for on-line reward optimization. In *ICML*, pages 956–963. DOI: 10.1145/1102351.1102472. 151

Shuai Wang, Zhiyuan Chen, and Bing Liu, (2016). Mining aspect-specific opinion using a holistic lifelong topic model. In *WWW*. DOI: 10.1145/2872427.2883086. 116

Christopher J. C. H. Watkins and Peter Dayan, (1992). Q-learning. In *Machine Learning*. DOI: 10.1007/bf00992698. 144

Xing Wei and W. Bruce Croft, (2006). LDA-based document models for ad hoc retrieval. In *SIGIR*, pages 178–185. DOI: 10.1145/1148170.1148204. 7, 91, 94, 109, 155

Peter Welinder, Steve Branson, Takeshi Mita, Catherine Wah, Florian Schroff, Serge Belongie, and Pietro Perona, (2010). Caltech-UCSD birds 200. 135 91

Robert West, Evgeniy Gabrilovich, Kevin Murphy, Shaohua Sun, Rahul Gupta, and Dekang Lin, (2014). Knowledge base completion via search-based question answering. In *WWW*. DOI: 10.1145/2566486.2568032. 73, 74

Marco Wiering and Martijn Van Otterlo. (2012). Reinforcement learning. *Adaptation, Learning, and Optimization*, 12. DOI: 10.1007/978-3-642-27645-3. 131

Aaron Wilson, Alan Fern, Soumya Ray, and Prasad Tadepalli, (2007). Multi-task reinforcement learning: A hierarchical Bayesian approach. In *ICML*, pages 1015–1022. DOI: 10.1145/1273496.1273624. 32

Rui Xia, Jie Jiang, and Huihui He, (2017). Distantly supervised lifelong learning for large-scale social media sentiment analysis. *IEEE Transactions on Affective Computing*, 8(4):480–491. DOI: 10.1109/taffc.2017.2771234. 8, 14, 15, 16, 140, 142, 143, 146, 152

Pengtao Xie, Diyi Yang, and Eric P. Xing, (2015). Incorporating word correlation knowledge into topic modeling. In *NAACL-HLT*, pages 725–734. DOI: 10.3115/v1/n15-1074. 51

Chen Xing, Wei Wu, Yu Wu, Jie Liu, Yalou Huang, Ming Zhou, and Wei-Ying Ma, (2017). Topic aware neural response generation. In *AAAI*. 92

Hu Xu, Bing Liu, Lei Shu, and Philip Yu, (2018). Lifelong domain word embedding via meta-learning. In *Proc. of 27th International Joint Conference on Artificial Intelligence (IJCAI)*. 131

Ya Xue, Xuejun Liao, Lawrence Carin, and Balaji Krishnapuram, (2007). Multi-task learning for classification with Dirichlet process priors. *The Journal of Machine Learning Research*, 8, pages 35–63. 14, 16, 36, 51, 52, 53

Wei Yang, Wei Lu, and Vincent Zheng, (2017). A simple regularization-based algorithm for learning cross-domain word embeddings. In *EMNLP*. https://www.aclweb.org/antho logy/D17-1311 DOI: 10.18653/v1/d17-1312. 27, 28, 53

Jason Yosinski, Jeff Clune, Yoshua Bengio, and Hod Lipson, (2014). How transferable are features in deep neural networks? In *NIPS*, pages 3320–3328. 52

Kai Yu, Volker Tresp, and Anton Schwaighofer, (2005). Learning Gaussian processes from multiple tasks. In *ICML*, pages 1012–1019. DOI: 10.1145/1102351.1102479. 24

Shipeng Yu, Volker Tresp, and Kai Yu, (2007). Robust multi-task learning with T-processes. In *ICML*, pages 1103–1110. DOI: 10.1145/1273496.1273635. 26

Bianca Zadrozny, (2004). Learning and evaluating classifiers under sample selection bias. In *ICML*, page 114, ACM. DOI: 10.1145/1015330.1015425. 27

Matthew D. Zeiler, M. Ranzato, Rajat Monga, Min Mao, Kun Yang, Quoc Viet Le, Patrick Nguyen, Alan Senior, Vincent Vanhoucke, Jeffrey Dean, et al., (2013). On rectified linear units for speech processing. In *Acoustics, Speech and Signal Processing (ICASSP), IEEE International Conference on*, pages 3517–3521. DOI: 10.1109/icassp.2013.6638312. 47

Friedemann Zenke, Ben Poole, and Surya Ganguli, (2017). Continual learning through synaptic intelligence. In *International Conference on Machine Learning*, pages 3987–3995. 69

Yusen Zhan, Haitham Bou Ammar, and Matthew E. Taylor, (2017). Scalable lifelong reinforcement learning. *Pattern Recognition*, 72:407–418. DOI: 10.1016/j.patcog.2017.07.031. 58, 74

Zhanpeng Zhang, Ping Luo, Chen Change Loy, and Xiaoou Tang, (2014). Facial landmark detection by deep multi-task learning. In *ECCV*, pages 94–108. DOI: 10.1007/978-3-319-10599-4_7. 140

Wayne Xin Zhao, Jing Jiang, Hongfei Yan, and Xiaoming Li, (2010). Jointly modeling aspects and opinions with a MaxEnt-LDA hybrid. In *EMNLP*, pages 56–65. 30

Fengwei Zhou, Bin Wu, and Zhenguo Li.   Deep Meta-Learning: Learning to Learn in the Concept Space. In *arXiv preprint arXiv:1802.03596*, 2018. 91

George Kingsley Zipf, (1932). *Selected Papers of the Principle of Relative Frequency in Language.* Harvard University Press. DOI: 10.4159/harvard.9780674434929. 34 104

Authors' Biographies

ZHIYUAN CHEN

Zhiyuan Chen completed his Ph.D., titled "Lifelong Machine Learning for Topic Modeling and Classification", at the University of Illinois at Chicago under the direction of Professor Bing Liu. He joined Google in 2016. His research interests include machine learning, natural language processing, text mining, data mining, and auction algorithms. He has proposed several lifelong learning algorithms to automatically mine information from text documents, and published more than 15 full research papers in premier conferences such as KDD, ICML, ACL, WWW, IJCAI, and AAAI. He has also given three tutorials about lifelong machine learning at IJCAI-2015, KDD-2016, and EMNLP-2016. He has served as a PC member for many prestigious natural language processing, data mining, AI, and Web research conferences. In recognition of his academic contributions, he was awarded Fifty For The Future® Award from the Illinois Technology Foundation in 2015.

BING LIU

Bing Liu is a Distinguished Professor of Computer Science at the University of Illinois at Chicago. He received his Ph.D. in Artificial Intelligence from the University of Edinburgh. His research interests include lifelong machine learning, sentiment analysis and opinion mining, data mining, machine learning, and natural language processing. He has published extensively in top conferences and journals in these areas. Two of his papers have received 10-year Test-of-Time awards from KDD, the premier conference of data mining and data science. He has also authored three books: one on Web data mining and two on sentiment analysis. Some of his work has been widely reported in the popular press, including a front-page article in the *New York Times*. On professional services, he served as the Chair of ACM SIGKDD from 2013-2017, as program chair of many leading data mining related conferences, including KDD, ICDM, CIKM, WSDM, SDM, and PAKDD, as associate editor of many leading journals such as *TKDE*, *TKDD*, *TWEB*, and *DMKD*, and as area chair or senior PC member of numerous natural language processing, AI, Web research, and data mining conferences. He is a Fellow of the ACM, AAAI, and IEEE.

Printed in the United States
by Baker & Taylor Publisher Services